# GEOLOGY AT THE UNIVERSITY OF MANCHESTER

# GEOLOGY AT THE UNIVERSITY OF MANCHESTER

## A BRIEF HISTORY, 1851–2004

David Vaughan and Jack Zussman

Matador
9 Priory Business Park,
Wistow Road, Kibworth Beauchamp,
Leicestershire. LE8 0RX
Tel: 0116 279 2299
Email: books@troubador.co.uk
Web: www.troubador.co.uk/matador
Twitter: @matadorbooks

ISBN 978 1789017 106

British Library Cataloguing in Publication Data.
A catalogue record for this book is available from the British Library.

Printed and bound by CPI Group (UK) Ltd, Croydon, CR0 4YY
Typeset in 12pt Adobe Caslon Pro by Troubador Publishing Ltd, Leicester, UK

Matador is an imprint of Troubador Publishing Ltd

MIX
Paper from
responsible sources
FSC
www.fsc.org    FSC® C013604

# CONTENTS

# ACKNOWLEDGEMENTS

We thank the University of Manchester and the University Archivist (Dr James Peters) for providing access to the university archives; and the School of Earth and Environmental Science and the Head of School, Professor Kevin Taylor, for support including financial assistance. Financial support was also received from the University President and Vice Chancellor (Dame Nancy Rothwell), and from the Dean of Science and Engineering (Professor Martin Schroeder).

We thank all those who have contributed material for our illustrations. Specifically, thanks are due to the British Geological Survey for permission to use a portrait of O.T. Jones (Figure 4); to Buxton Museum and Art Gallery (Derbyshire County Council) for permission to use a picture of Boyd Dawkins' reconstructed study (Figure 15); to Professor Lyon for providing a picture of mass spectrometers in the Cosmochemistry and Isotope Geochemistry Centre (Figure 18) and to Airtask Group for an image of the BAe-146 research aircraft operated on behalf of NERC by the (Manchester) Centre for Atmospheric Sciences working with the NERC

National Centre for Atmospheric Sciences (Figure 19). Unfortunately, it has not always been possible to determine the original source of a picture and we apologise for any resulting inadvertent breaches of copyright on our part. We invite any readers who do have copyright concerns to contact us to discuss them.

Many of our fellow academics responded to requests for information, and the whole manuscript was reviewed by Dr John Pollard and Professor Michael Henderson. Of course, any errors or omissions are entirely the responsibility of the authors.

# CHAPTER 1

## INTRODUCTION

## PRECURSORS AND BEGINNINGS

For more than 160 years, geology has been taught and researched at what is now the University of Manchester. Interest in the subject in Manchester, the world's first great industrial city, can be traced back to the Manchester Literary and Philosophical Society, founded in 1781; and more specifically to the Manchester Society for the Promotion of Natural History, founded in 1821; and later the Manchester Geological Society, established in 1838. A new museum for housing the collections of the latter two societies was built in Peter Street from 1835. Mr William Crawford Williamson (of Scarborough) was appointed as the first Curator of the new museum.

The formal study and advanced teaching of geology began at Owens College, the precursor of the university, in 1851. At that time, Mr W. C. Williamson was appointed as Professor of Natural History, a post which included botany and zoology as well as geology. Williamson also taught human anatomy and physiology, and retained his post as Museum Curator. Remarkably, in parallel with his interests in natural history, by 1841 Williamson had qualified and set up a practice in medicine; he later became a surgeon at the Manchester Ear Institute and taught and researched in medicine. He was also active as one of the pioneers in giving evening classes and popular lectures on science to the general public of the Manchester area. Such lectures were very much part of 19th-century intellectual life as recorded, for example, in the discourses at the Royal Institution in London given by Michael Faraday.

## OUR AIMS IN THIS BOOK

The history of geology at Manchester University is one of recorded observations and ideas on the one hand, and people and personalities on the other. It offers an example of how the subject has evolved as a discipline taught in the classroom, laboratory and field, and evolved as a research area from simple observations to advanced methods of analysis and experiment. In the short account that follows, we trace this history of ideas and of people over the generations, bringing to life through records and anecdotes how a 'natural' science can evolve over time. We begin in 1851 and end in 2004 with the merger of the Victoria University of Manchester (VUM) and the University of Manchester Institute of Science and Technology (UMIST). This also coincides with the incorporation of the Department of Geology, latterly renamed Earth Sciences, into a much larger School of Earth, Atmospheric and Environmental Sciences. Although our account of the earlier years relies on the written record, Jack Zussman was first appointed at Manchester in 1952 and, apart from the years 1955–1956 and 1962–1967, spent his whole career until retirement in 1989 at Manchester. David Vaughan was appointed to a Chair in Mineralogy in 1987, remaining in Manchester until the present day. As a result, we are able to tell a much fuller story for the years from 1952 until 2004 (and indeed, until 2017) compared with the years from 1851 to 1951.

Our aim in this book is to provide both an account of the developments in geology at Manchester, and to convey some sense of the people involved in these developments. In the first major chapter (Chapter 2), we discuss the Heads of the Department, their personalities, and key developments associated with their years in office. Tied in with these are the various 'comings and goings' of academic staff recounted in Chapter 3. There follows a chapter on research, Chapter 4, dealing with the 'classical' methods employed in the early years, new laboratories and major equipment, and the emergence of research institutes and centres. Chapter 5 concerns the student (and staff!) 'experience', and hence teaching methods, courses, and student numbers. Chapter 6 records personal recollections written by a number of former students and staff. A last major chapter (Chapter 7) covers outputs and recognition, ranging from grants won, books published, awards and honours, to minerals named for Manchester geologists, presidencies of learned societies and other such recognition. A short final chapter (Chapter 8) is by way of an epilogue, briefly dealing with key people and with major developments after 2004.

In seeking insights into what life was like for students of geology (and their teachers) at Manchester, we have sought input from past students through contacting alumni, and have also talked with retired staff. Overall, our book aims to provide a flavour of the history of geology at Manchester University, with personal insights alongside the academic record. In no way can our account be comprehensive, given our limited available resources of information and the need to keep the total length reasonable. Some errors in the text are inevitable and are entirely the responsibility of the authors. Omissions are also inevitable, and we apologise in advance to those who may feel 'hard done by' in our brief history.

# CHAPTER 2

# HEADS OF DEPARTMENT

As one might expect, the successive Heads of the Geology Department (later renamed the Department of Earth Sciences) played key roles in its development, in the evolution of the teaching programme and in the direction of research. In this chapter, we describe each Head and briefly record their contributions.

## WILLIAM CRAWFORD WILLIAMSON FRS (1851-1874)

As noted above, in 1851, Williamson (see Figure 1) was appointed the first Professor of Natural History at what was to become the University of Manchester. His duties involved instruction in zoology and botany as well as geology. In 1872 geology was separated, and in 1879 zoology was split off, leaving him with the Chair of Botany, from which he resigned in 1892 after forty-one years of continuous service.

In his early boyhood, Williamson acquired a good knowledge of field natural history from his father, a well-known Yorkshire naturalist, and his father's friends, notably William Smith (1769–1839), the founder of modern stratigraphical geology, and John Phillips (1800–1874), Professor of Geology at Oxford. On leaving school, he was apprenticed as a medical student (1832) to Thomas Weddell, apothecary of Scarborough. In 1835, he received a call from the Manchester Natural History Society to the curatorship of their museum, whereupon Weddell generously cancelled his indentures. Williamson held this office for three years, undertaking especially geological research and publication, and was a frequent visitor at the Literary and Philosophical Society, where he met, among others, John Dalton (1766–1844). A famous story concerning Dalton and Williamson had the latter watching by Dalton's bedside during his final illness and, becoming alarmed at his condition, mixing some hot brandy and water, which Dalton drank quickly and to Williamson's relief was heard to exclaim, 'That's good stuff.'

In the summer of 1838, in order to raise funds for medical study, Williamson gave a course of six lectures on geology in various towns of Lancashire, Yorkshire and Durham. He pursued his medical studies in Manchester and at University College London, qualifying in 1841and returning to practise in Manchester. Ear troubles in his student days had interested him in that organ, and he was actively involved in the creation, in 1855, of the Manchester

Institute for Diseases of the Ear, becoming surgeon to it until 1870. He continued both as a consultant and in general medical practice until he was seventy years old.

Williamson's scientific research work, pursued in the midst of his other professorial and medical duties, was substantial. In geology, his early work on the zones of distribution of Mesozoic fossils (begun in 1834), and on the part played by microscopic organisms in the formation of marine deposits (1845), was pioneering. In zoology, his investigations of the development of the teeth and bones of fish (1842–1851), and on recent Foraminifera, a group on which he wrote a monograph, were important. His work introduced the new technique of using thin sections for their study. In botany, his work on the structure of fossil plants established British palaeobotany on a scientific basis. Indeed, fossil plants were amongst his earliest interests, but it was only in 1858 that he began the comprehensive study of the fossil plants of the coal measures that is his main claim to fame. His memoir *On the Organisation of the Fossil Plants of the Coal Measures* was given as the Bakerian Lecture at the Royal Society in 1877. He was elected a Fellow of the Royal Society in 1854 and won their Royal Medal in 1874, along with many other honours.

As noted in Chapter 1, Williamson was a very successful lecturer on popular science, adding to his income. For example, between 1874 and 1890 alone he gave, amongst others, at least three hundred lectures in connection with the Gilchrist Trust. In an age long before even 'lantern slides', let alone PowerPoint presentations, he drew and painted beautiful diagrams to illustrate his talks, some of which were published in magazines such as the *Popular*

*Science Review*. An example was an article on *Pyrrhonism* (Scepticism) *in Science* in the *Contemporary Review* of 1881. Also in the spirit of popular science, he wrote an autobiographical book entitled *Reminiscences of a Yorkshire Naturalist* which was published in 1896 (see Appendix B).

## SIR WILLIAM BOYD DAWKINS FRS (1874–1909)

The college became responsible for the museum in 1867, and Boyd Dawkins (see Figure 2), who had studied classics and natural sciences at Oxford, was appointed as its Curator in 1869. Geology separated from Natural History and Botany in 1872, and he was appointed first as Lecturer in Geology, and in 1874 as the first Professor of Geology and Palaeontology in Manchester. In 1873, the collections were moved from Quay Street to new premises in Oxford Road. Their removal caused some amusement among the public as a procession of stuffed animals, including an elephant and a giraffe, passed along Oxford Road. Boyd Dawkins continued as Curator of the museum and with providing evening courses which attracted about thirty students, and in 1884 started 'field lectures' for geology students. He recognised the importance of practical 'demonstrations' and set up two laboratories; one for British fossils and sedimentary and other rock specimens, mineral specimens and models of crystal morphology, and the second for teaching applied geology including engineering drawing and geological maps. Facilities for chemical and microscope studies of minerals were also provided, and use for teaching was also made of the collections in the museum. In 1888, the present-day main museum building fronting Oxford Road was opened. In 1874, although no degree structure had yet been established, there were four students attending mineralogy and twenty-one attending geology courses. Evening classes in geology were attended by thirty-three.

Before moving to Manchester, William Boyd Dawkins had been employed by the Geological Survey of Great Britain. His research work involved mapping in Kent and the Thames Valley. When based in Manchester he played a key role in discovering the Kent Coalfields, from boreholes associated with the proposed Channel Tunnel project mentioned below. Boyd Dawkins was active in studies and publications on British Pleistocene mammals and investigations of caves. He made a special study of the mammalia of the Tertiary period and of problems associated with early man. His work in these areas was dealt with in his two books, *Cave Hunting* (1874) and *Early Man in Britain* (1880), both of which have been described by critics as *very readable*.

Boyd Dawkins was also one of the pioneers of engineering geology in Britain, with much knowledge of industry through consulting work, particularly in relation to hydrogeological schemes. He was involved with the Manchester Ship Canal and became Chairman of the Channel Tunnel Geological Survey Committee in 1882; this particular project was discontinued through lack of finance. He was elected FRS in 1867. He was President of the Manchester Geological Society in 1874–1875, 1876–1877, and 1886–1887. Dawkins resigned from the Chair of Geology in 1909, receiving an honorary DSc from Manchester, and was knighted in 1919.

## SIR THOMAS HENRY HOLLAND FRS (1909–1919)

Thomas Holland (see Figure 3) graduated from what became Imperial College London in 1889, and was awarded a Berkeley Fellowship for postgraduate study at Owens College, Manchester; however, in 1890 he was offered and accepted the post of Assistant Superintendent at the Indian Geological Survey in Calcutta. Holland also became Curator of the Geological Museum and Laboratory (1890–1896) and the first part-time Lecturer in Geology at the Presidency College in Calcutta. He was appointed Director of the Geological Survey of India in 1903, and although he made major advances in administrative aspects of science and education he also contributed much to petrographic and chemical research, recognised by his election to Fellowship of the Royal Society in 1904.

In India, as well as the investigations of such events as landslip, dam breakage and earthquakes, T. H. Holland's geological interests were many, including coal, manganese ore, micas, garnets, laterites and petroleum. He was the first to describe the charnockite rocks near Calcutta, named after Job Charnock, the founder of that city. It has been recorded that when Job died and an expensive pure white marble was about to be ordered from Italy for the tombstone, Sir Thomas named the location of one equally pure white marble in India and insisted on its use instead.

Sir Thomas received the medal KCIE (Knight Commander of the Order of the Indian Empire) in 1908 for his service to the Geological Survey of India, and was also KCSI (Knight Commander of the Order of the

Star of India). In 1909 he was appointed Professor of Geology and Mineralogy and Head of Department at Manchester, where his main research interests were in mineralogy, petrology, geodesy and mineral deposits, and he introduced a new course in mineral economics. He was awarded the Bigsby Medal of the Geological Society of London in 1913. Sir Thomas was well known also for his active interest in archaeology and cave exploration. He was President of the Manchester Geological and Mining Society (1912–1914), and of the Institute of Mining Engineering, London in 1916. In wartime Manchester he was in charge of the University Officers' Training Corps, with the rank of Major.

During the war years 1916–1918, he returned to India to be President of the Indian Industrial Commission in 1916 and of the Munitions Board from 1917 to 1919, and was appointed to the Viceroy's Executive Council from 1920 to 1921. A four-volume series, *Provincial Geographies of India*, was edited by him. He resigned from Manchester University in 1919, but on leaving was appointed to its Advisory Council.

Sir Thomas then took up the position of Rector of Imperial College London (1922–1929). He was Vice President of the Royal Society (1924–1925), and President of the British Association (1928–1929), the Geological Society of London (1933–1934), and the Mineralogical Society of Great Britain and Ireland (1933–1936).

Finally, Sir Thomas became Principal and Vice Chancellor of the University of Edinburgh in 1929, was elected Fellow of the Royal Society of Edinburgh in 1930, and was its Vice President from 1932 to 1935. He received numerous honorary degrees and gave many presidential addresses on topics relating to higher education, natural resources and geology. He was a strong advocate of the concept of continental drift, and was author of *Mineral Sanctions as an Aid to International Security* (Oliver and Boyd, 1935; see Appendix B). He retired in 1944.

## OWEN THOMAS JONES FRS (1919–1930)

Owen Thomas (better known as 'O. T.') Jones (see Figure 4) spoke only Welsh until he entered Pencader Grammar School near Newcastle Emlyn, West Wales. A star pupil, his education saw him win a scholarship to what is now Aberystwyth University, where he obtained a first-class honours degree in physics, and on to Cambridge where he developed interests in geology and mineralogy. In 1894 he graduated in natural sciences, garnering numerous prizes during his time as a student in Cambridge. Realising the importance of fieldwork in geology, he then worked with the Geological Survey of Great Britain in mapping the western parts of the South Wales Coalfield. In addition to his 'official' survey work, he pursued his own research (in those days such work was self-financed), also directed towards the geology of Wales. In 1909, for example, he published an important paper on the geological structure of the Plynlimon area in Mid Wales. This work still remains the standard work on the Palaeozoic rocks of Mid Wales, and indeed of similar rocks worldwide. Leaving the survey in 1910, he was briefly a Lecturer and then became the Professor at his old college in Aberystwyth. He was a superb teacher and Director of Research, greatly admired by his colleagues and students. For two years he was the only member of staff in his department; his popularity as a Lecturer was sustained despite some of his classes starting at 7am. In 1919 he was persuaded to accept the Chair in Geology at Manchester, where he remained until 1930 when he became Woodwardian Professor of Geology at the University of Cambridge.

In a professional career of over fifty years, O. T. Jones undertook research in many areas of geology and mineralogy. These ranged from the lower Palaeozoic geology of Wales, to Silurian rocks in Britain and the USA, the Upper Towy and Llandovery District drainage system, and geophysical studies of the UK continental shelves. His interests in mineralogy focused on lead-zinc ores and particularly the lead-zinc mineralisation in northern Cardiganshire and western Montgomeryshire, which reached fruition with the 1922 publication by HM Stationery Office of the standard work on these mines (see Appendix B).

O. T. Jones served as President of both the Geological and Mineralogical Societies, and was elected a Fellow of the Royal Society and received its Royal Medal in 1956. The Geological Society awarded him both its Lyell and Wollaston Medals, whilst in 1958 he was also made an honorary LL.D by the University of Wales.

## SIR WILLIAM J. PUGH FRS
## (1931–1950)

William John Pugh (see Figure 5) graduated with the degree of BA in geography at the University College of

Wales, Aberystwyth, in 1914, but had become interested in geology through attending an ancillary course under Professor O. T. Jones, the Head of the Geology Department, and working with him as assistant on the fieldwork and mapping of the geology of the Aberystwyth District. A paper was published by them jointly in 1915.

The First World War, however, interrupted William Pugh's career. He served with the Royal Welsh Fusiliers from 1914, attached to General Staff of Army Headquarters in France and on the Rhine. He was twice mentioned in dispatches, received the OBE and the Croix de Guerre, and was demobilised with the rank of Major. Pugh's return to Wales coincided with O. T. Jones moving to take up the Chair of Geology at Manchester, and he was appointed in place of O. T. to the Chair of Geology at Aberystwyth in 1919, at the age of twenty-seven. In the years to 1931 he played a full part in teaching and achieved growth in the department, and was Dean of Science from 1929 to 1931. He continued energetically (in collaboration with O. T. Jones) with mapping the boundary between the Ordovician and Silurian geology of North Wales. He was awarded a DSc by the University of Wales in 1928.

When O. T. Jones moved from Manchester in 1931 to take up the Chair of Geology at Cambridge, it was time for Pugh to follow Jones again and move to be Professor of Geology and Director of the Laboratories at Manchester. Numbers of students and staff were small, but the range of geology required was broad so teaching loads were heavy, and Professor Pugh taught not only stratigraphy and palaeontology for honours students, but also made a large contribution to the more basic first year. He nevertheless continued research on the Ordovician/Silurian boundary in North Wales, and again jointly with O. T. Jones, studied the Ordovician rocks of the Builth-Llandrindod area in Central Wales, and they revised older maps and discovered at Builth a fossilised 450-million-year-old shoreline.

During Pugh's time in office the Palaeontology Laboratory was refurbished and a field excursion, usually of one week's duration, was provided in each of the three undergraduate years. Professor Pugh served the university as Dean of the Faculty of Science (1940–1941), Pro-Vice-Chancellor (1941–1943), Deputy Vice Chancellor (1943–1950), and Member of Court and Council. He served as Chairman of the University Joint Recruiting Board (1935–1947). Outside of the university he was a member of the Manchester Education Committee, and a Governor of the United Manchester Teaching Hospitals. He became a member of the Geological Survey Board (1940–1944 and 1946–1950) and served on the Council of the Geological Society of London three times, but for some reason declined the presidency. Professor Pugh was President of the geological section of the British Association for the Advancement of Science for 1948–1949. He was awarded the Murchison Medal of the Geological Society of London in 1952, and received honorary degrees of LL.D from the University of Wales and DSc from the University of Nottingham. He resigned to become Director of the Geological Survey of Great Britain and the Museum of Practical Geology (1951–1960). Professor Pugh was elected Fellow of the Royal Society in 1951 and knighted in 1956.

## W. A. DEER FRS
## (1950–1961)

In 1950, Alex Deer (see Figure 6), then Senior Lecturer in Petrology at Cambridge, was appointed Professor of Geology and Director of Laboratories, replacing W. J. Pugh. In fact, Deer was returning to Manchester where he was born and brought up, and to Manchester University where he had graduated in geology in 1932 and had been awarded a Beyer research studentship to continue working in Manchester. However, he moved to take up a research studentship award at Cambridge University and went on to obtain his PhD degree there in 1937. During the course of his studentship, he joined the East Greenland Expedition (1935–1936) led by L. R. Wager, which had eleven participants, seven of whom, including Wager, his wife Phyllis, and Deer, stayed on through the long Arctic winter. This led to the classic work by Wager and Deer on the Skaergaard Layered Intrusion, which revolutionised igneous petrology. Deer also led an expedition to Baffin Island, Canadian Arctic, with Chris Brasher, and he was co-leader with Wager of another Greenland Expedition in 1953 in which two other Manchester academics participated (G. D. Nicholls and P. E. Brown). Deer resigned in 1961 to take up the appointment as the first Professor of Mineralogy and Petrology in Cambridge.

Deer's important publications included the Skaergaard memoir with Wager (1939), and over many years from 1962 the series of reference books *Rock-Forming Minerals* (first and second editions), and the textbook *An Introduction to the Rock-Forming Minerals* (three editions), most of which were co-authored with R. A. Howie and J. Zussman, who had both been appointed to the staff in Manchester during Deer's headship.

Professor Deer was elected FRS in 1962, became Master of Trinity Hall, Cambridge (1966–1975), President of the Mineralogical Society (1967–1970), a Trustee of the British Museum (Natural History; 1967–1975), President of the Geological Society (1970–1972) and Vice Chancellor of Cambridge University (1971–1973). In the Second World War, Deer served in the Royal Engineers (1940–1946) in Northern Ireland, the Middle East, Burma, and Kohima, North-East India, and left the Army with the rank of Lieutenant Colonel. The mineral deerite was named for him by Dr Stuart Agrell in 1960.

## E. A. VINCENT
## (1961–1966)

In 1961, Dr E. A. ('David') Vincent (see Figure 7), mineralogist, petrologist, geochemist, Reader in Mineralogy, University of Oxford, was appointed Professor of Geology and Head of Department. He was old enough to have been drafted to undertake wartime work which had involved the analysis and testing of explosives. David's career closely followed that of his mentor, Laurence ('Bill') Wager, by whom he was taught as an undergraduate at Reading, and under whose supervision he went on to study for a PhD on the Tertiary rocks of East Greenland during Wager's time at Durham. David then became a Demonstrator (1951) and later Reader in Wager's department in Oxford. Following his brief headship in Manchester, David was appointed to the Oxford Chair following Wager's untimely death.

His main research studies were on sulphide and oxide minerals, particularly the iron and titanium oxides of the Skaergaard Intrusion, East Greenland, and benefited from his experience and skills in reflected light microscopy. An expert also in chemical analysis by classical methods, Vincent's short time in Manchester saw the installation of an electron probe microanalyser, a laboratory for neutron activation analysis, and an electron microscope. He also made good use of the Government's decision based on the Robbins Report (1963) to finance the expansion of the universities in size and number, making a considerable number of new staff appointments (see below). Professor Vincent resigned in 1966 to return to Oxford to the Chair of Geology and Mineralogy. He was President of the Mineralogical Society 1974–1975, and the mineral vincentite was named after him by Dr Eugen Stumpfl in 1974. Strength in ore mineralogy and reflected light microscopy was added later by the appointment of E. Stumpfl, and later by A. C. Dunham, R. A. D. Pattrick and D. J. Vaughan, who also co-authored (with J. R. Craig) the main textbook in the field (see Appendix B).

## JACK ZUSSMAN
### (1967–1989)

Jack Zussman was one of the longest serving Heads of the Department of Geology (twenty-two years). He has an interesting history, being a fifteen-year-old at the start of World War II when he was evacuated with his school in East London to a small town in Somerset to escape the Blitz. On finishing school he volunteered to join the Royal Navy and saw service as the radar mechanic aboard a destroyer, including time spent escorting convoys on the infamous route to Arctic Russia. At the end of the war, he took up a promised place at Downing College, and graduated from Cambridge University with an honours degree in physics. He then went on to PhD research at the famous Cavendish Laboratories. At that time, the Director of the Cavendish was Laurence Bragg, and Jack's fellow researchers included Francis Crick and James Watson, discoverers of the crystal structure of DNA. Jack himself used the state-of-the-art facilities available in Cambridge to determine the crystal structures of an amino-acid and a nucleoside derivative.

In 1952, Professor Deer, then Head of Geology at Manchester, had the foresight to create a new junior post in mineralogy and crystallography with a view to appointing an X-ray crystallography specialist to work on the crystal structures of minerals. Although Jack had worked only on biomolecules, he had all the skills required for this task, and was duly appointed to the new position. He set up an X-ray laboratory and began his new line of research, working on the chemistry and crystal structures of minerals, particularly members of the serpentine and amphibole groups. Although this was the first use of X-ray diffraction in the Geology Department, it had been in use since the 1920s in the Physics Department in Manchester in work by W. L. Bragg and G. B. Brown (olivine), W. H. Taylor (aluminosilicates and feldspars) and B. E. Warren (pyroxenes and amphiboles) and others, to determine the structures of many common and important minerals.

It was at about this time that Alex Deer and Jack were joined in Manchester by Bob Howie, and a few years later the three of them began working together on a book entitled *Rock-Forming Minerals*. The book turned into five volumes, first published in 1962. In 1966, a condensed single-volume version (*An Introduction to the Rock-Forming Minerals*) was published for the student market. Over subsequent decades, second editions of these books have been published, and a third edition of the single-volume version. Their writing has proved a lifetime of work for the three original authors (in some cases joined by invited co-authors). Although Jack retired in 1989, he continued for many years working on the second editions and a third edition of *An Introduction to the Rock-Forming Minerals*.

Other high points in Jack's career include leading a team from Manchester Geology in studying samples returned to Earth by the American-manned landings of the Apollo lunar missions, and leading the Manchester bid in the Earth Science Review which led to an unprecedented expansion of the department in 1988. Jack also served with distinction as Dean of Science, which included a difficult time in the early '80s coping with the consequences of Government cuts in university funding. He has received many honours in recognition of his work, ranging from the naming of a mineral in his honour (zussmanite, see Appendix A) to designation of newly refurbished teaching laboratories in the Williamson Building as the Zussman Laboratories.

Probably through Jack's involvement in physics, chemistry and geological sciences he was a valued member of University Grants Committee (UGC), Royal Society, British Crystallographic Association and British Council committees. He chaired the Natural Environment Research Council's (NERC's) Geological Sciences Research Grants Committee for four years.

## CHARLES CURTIS OBE
## (1989-1992)

The retirement of Professor W. S. MacKenzie provided the opportunity to make a professorial appointment at a critical time for geology in UK universities. His Chair was advertised and in the event two professorial appointments were made, the second in the light of the impending retirement of Jack Zussman. Charles Curtis and David Vaughan were appointed at the same time in September 1987 and took up their posts in the spring of 1988. These appointments were made in advance of the major reorganisation of UK Geology Departments brought about following the Earth Science Review.

Charles Curtis was already the holder of a Chair, at Sheffield University which is where he had spent most of his academic career until 1988. His appointment at Manchester was as Professor of Geochemistry, and at the time of his appointment he was already renowned as a sedimentary geochemist undertaking both fundamental research and work of interest to the petroleum industry. Just before coming to Manchester he had spent a sabbatical year at the BP Research Laboratories in Sunbury. On arrival in Manchester, he was able to attract substantial research funding from the petroleum industry.

Charles, in his later years at Manchester, got involved in other areas of research and teaching, leading bids to establish environmental science programmes, including the University of Manchester Environment Centre (UMEC) which he directed. He also served the university as Research Dean of the Faculty of Science, encouraging colleagues to compete for funding both nationally and internationally. In his later years at Manchester, before taking early retirement, he became heavily involved in both research and policymaking relevant to the nuclear industry. He served as Chair of the (UK) Radioactive Waste Management Committee, and Head of Research and Development Strategy of the Nuclear Decommissioning Authority. His considerable abilities and contributions have been recognised by his election as President of the Geological Society, and by the award of the honour of Officer of the British Empire (OBE).

## DAVID J. VAUGHAN FRSC (1993–1997 AND 2001–2003)

David Vaughan is the only Professor to have served two terms as Head of Department at Manchester (a total of six years). He was invited to apply for the Chair ahead of the Earth Science Review (see Chapter 4), so as to strengthen the case to be made to the UGC for expanding and revitalising the Manchester Department. Following on from an active role in the Manchester bid, he led the initiative that resulted in establishing a Geoscience Research Institute and, a decade later, led the application for UGC (NERC) funding to set up the Williamson

Research Centre (WRC) for Molecular Environmental Science. He served for ten years as the first Director of the WRC (2001–2011).

Educated at the Universities of London and Oxford, from which he holds the degree of DSc, he worked at the Canada Centre for Minerals and Energy Technology, Ottawa (1970), the Massachusetts Institute of Technology (1971–1974), and the University of Aston in Birmingham (1974–1988). He has also been a Visiting Professor at the Virginia Polytechnic Institute and State University, the University of Florence, and Liverpool University, and is an Honorary Research Fellow at the Natural History Museum, London.

The research interests of David Vaughan centre on fundamental studies of minerals, particularly metal sulphides and oxides, using advanced analytical, spectroscopic and imaging techniques; molecular scale studies of mineral surfaces including interactions with microbial species, and applications of such studies to problems of Earth resources (including mineral extraction technologies) and the environment. His earlier contributions were in economic geology, ore mineralogy, and ore microscopy where he is co-author of the most widely used textbook in the field (see Appendix 2), and in sulphide mineral chemistry where he has also co-authored or edited major texts. In later years he pioneered the emerging field of 'environmental mineralogy' and also, through the establishment of the WRC, the field of molecular environmental science. Environmental work has included studies of the impact of acid mine drainage, atmospheric mineral dusts, nanoparticles, radioactive waste materials and depleted uranium munitions.

In keeping with the traditions of the Manchester Department, David Vaughan has had a major interest in promoting the geosciences through lectures (see below) and book writing. He is co-author of the premier undergraduate text on Earth resources, and an introductory book on minerals aimed at a general audience (see Appendix 2).

In 2003–2004 he was Mineralogical Society of America Distinguished Lecturer. In 2005, he received the Royal Society of Chemistry Award in Geochemistry and, in 2006, the Schlumberger Medal of the Mineralogical Society of Great Britain and Ireland. He is one of the few UK scientists elected (in 2007) as a Geochemical Fellow of the Geochemical Society and of the European Association of Geochemistry. In 2007–2008 he was Mineralogical Society Distinguished Lecturer, in 2010–2012 Councillor of the Geological Society, and became President of the Mineralogical Society of America in 2014. He has the unprecedented distinction of having served as President of three Mineralogical Societies; of Great Britain and Ireland, America, and Europe (the European Mineralogical Union). He was also honoured by the Royal Society of Canada through his (2016) election as a Fellow (FRSC), in recognition of his research on subjects of importance to Canadian science. The sulphide mineral vaughanite, from Hemlo, Canada, was named for David Vaughan in recognition of his contributions to mineralogy.

# ERNEST H. RUTTER
# (1998-2000)

Ernie Rutter moved from a readership at Imperial College London in 1989, as part of the implementation of the Earth Science Review (see Chapter 4). The costs of moving both home and laboratory were paid in full by the University Grants Committee (UGC). With the move, Ernie retained the title of Reader in Structural Geology, but was promoted to Professor only a few years later. Most of Ernie's research has been concerned with the experimental deformation of rocks under simulated geological conditions, in order to help interpret natural processes (such as in mountain building) leading to the flow and fracture of rocks. Apart from the intrinsic scientific interest of rock deformation processes, the results of such studies also support large-scale geodynamic modelling of the evolution of the Earth's crust and mantle. Rock mechanics also underpins the understanding of rock behaviour under engineering conditions, with implications for the stability of slopes, excavations and boreholes, and hence the exploitation of natural resources. In addition to laboratory-based experimentation, through his primary training as a geologist, Ernie carried out researches in various areas of field-based geoscience, often being able to bring an experimentalist's approach to illuminate essentially field-geological problems. Ernie has overseen the development of a substantial laboratory for experimental rock deformation, equipped with nine mechanical testing machines of various different types, able to deform rocks at pressures up to 500 MPa, temperatures of 1,200°C, and with controlled pore fluid pressures. These facilities are supported by the necessary additional equipment for sample fabrication and characterisation before and after testing. By 2016, the laboratory had supported the PhD and MSc projects of respectively thirty-three and twenty-nine students, most of whom went on to develop highly successful research or industrial careers.

Studies undertaken by Ernie and his students and collaborators have included work on pressure solution as a deformation mechanism, plasticity and grain-size-sensitive flow in quartz and calcite rocks, flow of partially molten granitoids, and fabrics of deformed rocks. Deformation-metamorphism interrelationships during rock deformation, and deformation and dehydration of serpentinites and micaceous rocks have also been studied. Physical properties of rocks, particularly the seismic properties of lower crustal rocks and seismic and permeability properties of fault rocks, have been measured. Field studies of the internal geometry of the Ivrea-Verbano zone, Northern Italy (one of the best exposed sections of the lower continental crust), and of the internal structure of the Carboneras Fault (a new class of transform fault) in South-Eastern Spain, have complemented laboratory work which has included acoustic properties of the fault rocks and their protoliths, deformation of porous sandstones, and fracture, friction and permeability of gas shales.

In recognition of his contributions to science, Ernie has been awarded the Wollaston Fund of the Geological Society (1993) and its Lyell Medal (1999), the Néel Medal of the European Geoscience Union (2011) and Fellowship of the American Geophysical Union (2004). Later years have seen his expertise widely sought in relation to the (currently controversial) method of extracting oil and gas known as 'fracking'.

## RICHARD A. D. PATTRICK
## (2003-2006)

Richard Pattrick joined the department in 1980, coming directly from his PhD studies at Strathclyde University. In effect, he replaced Ansel Dunham so as to provide expertise in economic geology and ore mineralogy. He served as Head of Department during the transition associated with the merger of Manchester and UMIST when the Department of Earth Sciences became a School of Earth, Atmospheric and Environmental Sciences through the acquisition of the Atmospheric Physics Group formerly based in UMIST.

Richard's earlier research was concerned with British mineral deposits, on which he co-edited a book (see Appendix 2), and on the mineral chemistry of the complex tetrahedrite group of sulphide minerals. Although retaining interests in these topics, his later work became much more wide-ranging. In the mid 1980s he worked with David Vaughan and David Garner (then Head of Chemistry) in the application of synchrotron methods (particularly X-ray absorption spectroscopy or XAS) to sulphide minerals including the tetrahedrites, and on the structural development of amorphous metal sulphide precipitates. At the same time, his work on mineral deposits benefited from collaborations with the Manchester isotope geochemists, particularly in studying the noble gases trapped in fluid inclusions representative of mineralising fluids, and in developing a Xe-Xe geochronometer for telluride minerals.

With the establishment of the Williamson Research Centre for Molecular Environmental Science, Richard became involved in a wide range of projects in collaboration with the geomicrobiologists. These included work on the synthesis of bio-nanominerals with interesting magnetic and semiconducting properties, where he provided the techniques for characterisation using the synchrotron. Fundamental work on mineral chemistry and geochemistry latterly became directed towards more applied contributions, ranging from the behaviour of toxic metals (especially selenium) in the environment, geological disposal of nuclear wastes, alpha particle damage to irradiated mineral structures, and metal sorption on sulphide mineral surfaces in the context of mineral processing by froth flotation.

Richard Pattrick proved a very able administrator during a time of major changes. As well as serving as Head of Department (School), he earlier served as Graduate Dean in the Faculty of Science and Engineering, and Head of the Graduate School covering all science, engineering and biomedical sciences. The knowledge he acquired of synchrotron methods led to his serving as a member, and later Chair, of key committees advising on the development and review of the scientific programme at the UK Synchrotron Facility (Diamond Light Source). He also served as a very effective President of the Mineralogical Society.

# CHAPTER 3

## ACADEMIC STAFF: THE COMINGS AND GOINGS

Geology at the University of Manchester has been a story of the 'comings and goings' of academic staff. Many have seen out all, or a large part, of their careers in the department. For others it has been a stepping stone to a more senior academic post or position elsewhere, such as in a Geological Survey.

In the early years, there was only 'the Professor', joined later by a junior staff member to help with the teaching and with the research undertaken by the Professor. In Manchester, as noted above, William Crawford Williamson was active in many different areas, and eventually he was joined by William Boyd Dawkins who was appointed as a Lecturer in Geology in 1870. Subsequently also appointed to assist Williamson in his research was the palaeobotanist Thomas Hick (1886–1896). In 1880, the year that 'the Victoria University of Manchester' received its Royal Charter, the university prospectus shows that geology and palaeontology were taught by Professor Boyd Dawkins. In these early years, mineralogy (lectures and practical work) was taught in the Chemistry Department by Mr Thorpe (1869–1870) and by Dr Charles A. Burghardt (1871–1898). The Mr Thorpe mentioned here began working in 1863 as an assistant to Henry Roscoe, the distinguished Professor of Chemistry at Owens College, and went on himself to become Sir Edward Thorpe, another very eminent chemist.

The Honours School in Geology and Mineralogy was established in 1881, and in 1887, new laboratories (the Beyer Laboratories) were opened to provide facilities for the Departments of Botany, Geology and Zoology. At the same time, the new Manchester Museum opened at its present site on Oxford Road. Also at this time, P. F. Kendall, assistant to Boyd Dawkins, was appointed Demonstrator and Assistant Lecturer but resigned after two years. Bernard Hobson was then appointed as Assistant Lecturer in Petrology, promoted to Lecturer in 1899 and served until 1910.

With the start of the new century, the University of Manchester became an independent body, having formerly been part of a federation with Leeds and Liverpool Universities. At this time, arguably the most famous person to work in the geosciences at Manchester University was Marie Stopes, appointed as Assistant Lecturer in Palaeobotany (in the Department of Botany; 1904–1907) and after a period researching in Japan, returning to Manchester as a Lecturer (1909–1911). Her research was mainly on fossil plants in coals (coal macerals), but she also studied fossil angiosperms, glossopteris and cycads. However, she is particularly famous for being a leading campaigner for birth control and women's suffrage. She was also the first woman to hold a teaching post in the Faculty of Science at the University of Manchester, and wrote a popular book on ancient plants (see Appendix B). Figure 8 shows Stopes sitting at her microscope in 1905.

In 1906, H. G. A. (George) Hickling was appointed Demonstrator and Assistant Lecturer. Hickling gained a Manchester honours BSc degree in geology in 1905 and then started to research on Permian footprints in Britain, and to work on his highly regarded book *Geology: Chapters of Earth History*. In the following year he was appointed Assistant Lecturer in Stratigraphy and Palaeontology, thereby allowing more time for Professor Boyd Dawkins to teach applied geology. Hickling's research led to the award of an MSc degree in 1909 and a DSc in 1910. During and after World War I he carried out extensive research on the structure and stratigraphy of the South-East Lancashire Coalfield, and became an expert on the micro-petrography and geochemistry of coal, publishing *The Geological History of Coal*. He also described the anatomy of calamite cones and collected many fossil plants from the Old Red Sandstone, as a result of which the plant *Hicklingia* was named after him. Hickling left Manchester in 1920 to the Chair of Geology at Armstrong College, later known as King's College, part of Newcastle University. He was awarded the Murchison Medal by the Geological Society of London in 1934, and became a Fellow of the Royal Society in 1936.

Another Manchester Geology graduate with a remarkable career was D. M. S. Watson. He graduated in 1907 with a first-class BSc in geology, having already published in the journal of the Microscopical Society his research on the 'fern' Synangium from the Lower Coal Measures, and an important paper with Marie Stopes on coal balls in 1907. He was awarded a Beyer Fellowship in 1908, and an MSc degree and appointment as Demonstrator in Geology at Manchester in 1909. His interests then turned to vertebrate palaeontology, using material borrowed mainly from the Natural History Museum in London. He moved to University College London in 1912, as Lecturer in Vertebrate Zoology. During World War I, Watson, holding the rank of Captain in the Royal Air Force, became involved in matters concerning airship and balloon fabrics. After his return to UCL he was appointed to the Jodrell Chair of Zoology and Comparative Anatomy in 1921. Amongst numerous honours, Watson was elected a Fellow of the Royal Society in 1922.

The museum building and bridge over Coupland Street were completed in 1909, and the biology and geology combined honours degree was first offered in 1912. A reflection of the strong presence of mining in the curriculum was indicated by the department being known for some time as the Geological and Mining Department. This also led to the Manchester Geological Society (established 1838) being renamed the Manchester Society for Geology and Mining, and the eventual formation (1925) of a separate society named the Manchester

Geological Association (MGA). The MGA remains a very active body, with a membership from both within and outside the university, as seen by the tradition of alternately electing an MGA President from members within and from outside the department staff. During the early 20th century, physical geography was another prominent area, as reflected in the appointment of A. Jowett, whose fields of interest were geomorphology and glacial geology, as Assistant Lecturer (1912–1914).

The work of the university was greatly affected during the period 1914–1918, the years of the First World War. At this time (1914), F. G. Percival was appointed Junior Assistant Lecturer and Demonstrator, but he left to serve as a 2nd Lieutenant in the Royal Garrison Artillery, returning in 1919. Shortly after returning, he resigned to take up a post in Calcutta. During much of the time from 1916 to 1918, although T. H. Holland was officially Head of Department, he was actually employed on Government service in India and H. G. A. Hickling served as Acting Head of Department. Miss Margaret March was appointed as a temporary assistant at this time and was involved in research on the Lancashire Coalfield, particularly on coal measure fossils. The year 1919 also saw a new undergraduate course on mineral economics and mining become available.

The next few years also saw a series of new and replacement appointments, with G. K. Charlesworth, whose expertise was in physical geography and glacial geology, being replaced as a Lecturer in Geology by G. Andrew (1921). N. T. Williams, who had returned from military service in France, was given responsibility for lectures in mining, mine surveying and planning. Miss M. Lindsey taught palaeontology until the appointment of Dr S. H. Straw in 1923, a palaeontologist whose interests included Liassic corals, fossil fish, and the Silurian in Central Wales. In addition to his own research and teaching, Straw acted as assistant to three Heads of Department (Jones, Pugh and Deer), and was Acting Head during the gap between the resignation of O. T. Jones and the appointment of Pugh in 1931. He eventually retired as a Reader in 1958. Also at about this time, Mr Serge Tomkeieff (later Professor and Head of Geology at Newcastle University, presumed to be a Visiting Lecturer at Manchester, although that is not clear from records) gave a special course on determinative mineralogy. Dr R. W. Palmer, a Manchester graduate (1908–1912), was appointed as Senior Lecturer but sadly died soon after, and Mr Edgar Morton (a Consultant Geologist and Manchester graduate, 1919–1922) was appointed part-time to teach geology to engineering students. This activity proved highly successful with large numbers of students, and was still going strong in 1961 when C. W. Isherwood was appointed as a part-time Special Lecturer in Engineering Geology to assist Edgar Morton. The latter became a widely respected consultant across a wide range of geological problems including those concerned with dams and reservoirs, building foundations, land stability, and the exploitation of mineral resources. He was promoted to Reader in Applied Geology in 1963 and retired in 1966.

A very important figure from this era, although not based in the Geology Department, was Sir Henry Alexander Miers FRS. In 1915, Miers, an eminent mineralogist, was appointed as Vice Chancellor of the University of

Manchester. He was well known for his book *Mineralogy: An Introduction to the Scientific Study of Minerals*, published in 1902 while he was Waynflete Professor of Mineralogy at the University of Oxford (1895–1908), prior to which he had been employed at the British Natural History Museum (1882–1895). He left Oxford to take a position as Principal of the University of London until he came to Manchester seven years later in 1915. Miers' valuable contributions to mineralogy and university administration were recognised in many ways: Fellowship of the Royal Society (1896), a knighthood (1912), President of the Mineralogical Society of the UK and Ireland (1904–1909) and editor of its journal for nine years, and the Wollaston Medal of the Geological Society of London. He also received honorary doctorates from six universities. When Miers retired as Vice Chancellor in 1926, he was appointed as an Honorary Professor of Crystallography.

Another scientist from this era who initially worked as an assistant to Sir Henry Miers, and who was appointed to a new lectureship in Crystallography when Miers retired as Vice Chancellor, was H. E. Buckley. Loosely based in the Physics Department, but effectively an independent member of the Faculty of Science, Buckley became a leading authority on the morphology of crystals and on crystal growth. His book *Crystal Growth* (1951) was, at that time, the standard text on the subject. Buckley was promoted to Senior Lecturer (1937) and Reader (1952). There was little interaction between Buckley and the Department of Physics where, as described further below, pioneering work on the crystal structures of minerals was being undertaken by Nobel Laureate W. L. Bragg and a team including W.

H. Taylor, who determined the structures of many feldspar minerals.

In 1926, G. Andrew was replaced by Dr M. Hodge, who was interested in sediments from the north-west of England, particularly the Carboniferous coal seams of Lancashire and Permian Yellow Sands. Hodge, in his turn, was replaced by S. R. Nockolds who had been an undergraduate at Manchester. However, Nockolds moved to Cambridge for his PhD research in geochemistry and igneous petrology, where he was later appointed Lecturer and then Reader in Geochemistry. At Cambridge, he became a leading authority in petrology (elected FRS in 1959), and known to generations of students as co-author of the textbook *Petrology for Students* (1978; reissued 2010). Another geologist who was to become a noted authority on mineralogy and petrology appointed to a lectureship at Manchester was S. O. Agrell. He served from 1939 until 1949 before moving to Cambridge. His work in Manchester included studies of pyrometamorphosed basalts and industrial slags. In 1969, he was one of the leading scientists in the study of rocks returned by the Apollo manned missions to the Moon (see Chapter 4). Agrell had the reputation of being a genius at petrographic work with the optical microscope and at recognising new minerals.

During the Second World War, although teaching and research did continue at the university, developments were restricted by the priorities of war service and limited resources, with these limitations extending for some years after 1945. However, in 1949, three new Assistant Lecturers were appointed. With their fields of research interest, they were D. H. Griffiths (magnetic properties of rocks), J. D. Lawson (geology of Ludlow and Wenlock rocks), and J. H. Whittaker (petrology of rocks from Greenland). Also, G. W. Tyrrell, an eminent petrologist recently retired from Glasgow University, and author of *Principles of Petrology*, was hired for just two terms as an Assistant Lecturer. At this time, Dr G. D. Nicholls (a petrologist) was hired as a replacement for Stuart Agrell. As well as his research on the petrology of the Builth Volcanic Series, he set up laboratories for optical spectroscopy and, later, spark source mass spectrometry for trace element determinations in rocks. He also collaborated with the marine scientists from the Woods Hole Laboratories (Massachusetts, USA) in studies of Mid-Atlantic ocean floor basalts and glasses.

Probably not just by coincidence, the appointment of D. H. (Don) Griffiths took place at the same time as important developments in physics at Manchester. In the Physics Department, Professor P. M. S. Blackett FRS was moving his field of research away from nuclear physics to rock magnetism and the nature and origin of the Earth's magnetic field. Together with two other junior members of his department, S. K. Runcorn and J. A. Clegg, a research group on rock magnetism was formed by Blackett to work on improving the design of magnetometers and to make measurements on the strength and orientation of magnetisation in rocks from areas as diverse as Yorkshire, North Wales and India. Their measurements led to the conclusion that the Earth's magnetic field has undergone repeated and relatively rapid changes of polarity through the planet's history, and also provided evidence for what was then the much-disputed theory of continental drift.

In turn, this led a decade later to what is now regarded as a revolution in the geological sciences with the universal acceptance of the theory of plate tectonics.

Over the years 1950–1953 there were several changes of staff and new appointments. Dr I. M. Simpson (a stratigrapher) replaced J. H. Whittaker, and F. M. (Fred) Broadhurst, a palaeontologist, was appointed as Assistant Lecturer in 1951, replacing Lawson. Morven Simpson's research was concerned with stratigraphy at localities in Northern England (Pleistocene of the Manchester region and limestones of North Derbyshire being examples). He often led field courses and wrote a book on fieldwork in geology co-authored with Fred Broadhurst (see Appendix B). Morven's family suggest the existence of a 'geology gene'. His father (John Baird Simpson) was an eminent geologist who worked with the Scottish Geological Survey producing some of the first maps of the Western Highlands, and his son, who gained a geology PhD at Sheffield, went on to hold the Schlumberger Chair of Energy Management at Aberdeen University.

Fred Broadhurst had an interesting history. Immediately after the war, he contributed to the rebuilding effort by working in the coal mines of Lancashire as a so-called 'Bevin Boy'. This was an alternative way of contributing for those not wishing to join the military or other directed employment. It was in the coal mines that Fred acquired his lifelong passion for geology. He also had a great gift of enthusing generations of Manchester students with this passion. On meeting former students, the first question many would ask current staff or students is 'How is Fred?' He retired in 1990 after forty years of service as Lecturer, then Senior Lecturer in Geology. Famously, on one of his first-year field trips to Robin Hood's Bay, the party discovered a complete skeleton of a fossil Plesiosaurus which was later excavated and brought back to Manchester and exhibited. Fred's early research was on the palaeoecology of the Lancashire Coal Measures, and later work was concerned mainly with Dinantian reefs of the Peak District. He received the Silver Medal of the Liverpool Geological Society and the John Phillips Medal of the Yorkshire Geological Society. He was also very much involved in teaching and enthusing the general public about geology and initiated many extramural courses; this work was recognised by the National Institute of Adult Continuing Education Award to Fred as Adult Tutor of the Year in the North-West. The year 1950 saw two other appointments in palaeontology: Dr Mary Calder, a palaeobotanist, as a Senior Lecturer in both the Geology and Botany Departments, and C. H. Holland as Assistant Lecturer. Charles Holland, who was an undergraduate and postgraduate in Manchester, went on to a lectureship at Bedford College, London in 1953, and in 1966, to the Chair at Trinity College Dublin, from which he retired in 1993. He was President of the Geological Society of London from 1984–1986. As well as these appointments, in 1952 Dr Jack Zussman was appointed to a new post of Assistant Lecturer in Mineralogy and Crystallography (see Chapter 2 for more details about Jack). In 1953, Dr R. A. Howie was appointed as Assistant Lecturer in Geology (Mineralogy and Petrology), replacing Holland. It was in 1956 that the now-famous book-writing trio of Deer, Howie and Zussman first got together under the leadership of Alex Deer. (Figure 9 shows the three of them, along with MacKenzie, Morton, Simpson and Isherwood in 1961.)

Bob Howie was appointed Assistant Lecturer in 1953, having completed his Cambridge PhD on charnockite rocks from India. He was promoted to Lecturer in 1956, and at about that time started his lifelong collaboration with Deer and Zussman on the above-mentioned books. Although the writing work was shared equally between the three authors, Bob was the internal and 'business manager' for this substantial project. In 1962 he left Manchester for a readership at King's College London and was later promoted to be Lyell Professor, a post he retained after various mergers led him to being located at Royal Holloway University of London. Another important contribution initiated during his time at Manchester was the publication known as *Mineralogical Abstracts*, an abstracting service directed specifically at the mineralogical community. Bob's work was recognised by numerous honours including honorary degrees and honorary fellowships, the award of the Geological Society Murchison Medal, and presidencies of the Mineralogical Society and the Gemmological Society. The mineral howieite was named for him by Stuart Agrell (see Table 2, Appendix A).

The late 1950s saw another appointment that would prove to have a major impact on the history of geology at Manchester. Dr W. S. MacKenzie (known to all as 'Mac') was initially appointed as an ICI Research Fellow (1956) and rapidly promoted to Lecturer, then Senior Lecturer, and to a Chair in Petrology in 1964. As recounted in detail in Chapter 4, Mac was largely responsible for bringing the techniques of experimental petrology to the UK, having himself learnt them from colleagues in the USA. Where Mac worked, at the famed Geophysical Laboratory of the Carnegie Institution in Washington, DC, the very high temperatures and pressures associated with the crystallisation or melting of rocks could be created in the laboratory. In Manchester, Mac set up a laboratory to pursue such work, for example in his studies of phase equilibria in the system nepheline-kalsilite-silica-$H_2O$ (known as petrogeny's residua system, and important for the study of feldspars and the understanding of the petrogenesis of a wide range of igneous rocks). Mac was the organiser of the NATO Advanced Study Institute on the Feldspars, a prestigious international meeting held in Manchester in 1972, and which attracted all of the 'big names' in feldspar research. The conference, which included a field excursion to Shap Fell in Cumbria, was a great success (helped by ten days of continuous sunshine in Manchester) and the papers presented were published by Manchester University Press in a large hard-copy volume (see Appendix B). This event was an indication of the esteem in which Mac was held internationally, as was the international attendance at a conference held in the department on his retirement in 1988 after thirty-three years of service. One other noteworthy interest of Mac's was the study of rocks in thin section using the polarising microscope, and the art of producing photographs that accurately record such observations. In co-authorship with the Departmental Superintendent Cyril Guilford and various others (Donaldson, Adams and Yardley), such photographs were published in a series beginning with an *Atlas of Rock-Forming Minerals in Thin Section*, followed by similar volumes on igneous, sedimentary and metamorphic rocks (see Appendix B).

In 1958, S. H. Straw retired after thirty-five years of service and two Assistant Lecturers were appointed, both in structural geology. One was Dr Robin Nicholson, who spent the whole of his career of nearly forty years in Manchester, retiring as a Senior Lecturer. Robin was described in his Geological Society obituary as *a complete field geologist who combined deep knowledge and rigour with energy and endurance*; qualities he applied over many years in areas such as Arctic Norway. The other appointee was Dr John Dewey, who spent only four years in Manchester before moving on to a lectureship in Cambridge, and then professorships at the State University of New York (Albany) and Durham University, as Chair and Head of Department at Oxford and, latterly, a Professor at the University of California, Davis. John Dewey was one of the leading figures in the plate tectonic revolution and the recipient of numerous awards and honours including election to the Royal Society.

As already noted, the period during which E. A. ('David') Vincent took over as Head of Department was one of expansion and development. In terms of people, replacements included Dr W. L. Brown, a mineralogist and crystallographer with special expertise on the feldspar minerals (Bill co-authored a much-praised book on feldspars with J. V. Smith) who was appointed Senior Lecturer when Jack Zussman resigned to take up the post of Reader in Mineralogy at Oxford. Dr John Wadsworth, an igneous petrologist, replaced Bob Howie who moved to a readership at King's College London. John worked under Professor Wager at Oxford for his doctorate, studying mainly layered basaltic intrusions, and maintained this interest throughout his career, extending it to volcanic rocks from Scotland to Réunion Island, and the Comoros Archipelago. New posts were created for Dr John Pollard (a palaeontologist later specialising in trace fossils, a topic of which he became one of the leading exponents, with many publications and a fine collection of specimens for teaching and research) and Dr John Esson (a geochemist). John Esson brought with him expertise in neutron activation and wet chemical analysis, and extended this to set up laboratories for atomic absorption ('AA') and X-ray fluorescence spectroscopy, which provided the departmental staff and students with invaluable analytical capabilities. Later, he managed a department service providing these and analytical services using AA to external clients, which was a useful source of additional revenue. Wadsworth, Pollard and Esson all came to Manchester from Oxford and were soon dubbed 'the Oxford Johns'. Also at this time, Dr J. M. ('Mike') Anketell, although a sedimentologist (working chiefly in Central Wales and later in Libya) rather than a structural geologist, was appointed to replace John Dewey. Drs Wadsworth, Pollard, Esson and Anketell all remained in Manchester until reaching retirement. A game of what some might call 'musical (professorial) chairs' between the Manchester and Oxford Geology Departments also took place in the '60s which saw Vincent as Reader in Oxford and Zussman as Senior Lecturer in Manchester, Vincent moving to become Professor at Manchester and Zussman to be Reader in Oxford, and then Vincent returning to Oxford as Professor in return for Zussman becoming Professor and Head in Manchester.

The 1960s were a very productive and successful period for geology at Manchester. The many new staff appointments of young active scientists in turn attracted funds for research and senior academics to Manchester. The research base was strengthened by senior staff setting up laboratories for experimental petrology (see Chapter 4) and also X-ray diffraction, electron microscopy and electron probe microanalysis to study the compositions and structures of both natural and synthetic minerals. A good example of the developments in staffing was the appointment in 1965 of W. S. Fyfe to a special Royal Society-funded Chair in Geochemistry. Bill Fyfe, a New Zealander, came to Manchester from a professorship at the University of California, Berkeley. He was already probably the world's foremost geochemist and attracted many research students and postdoctoral workers and the funds to support them. Bill was elected FRS in 1969 and left Manchester for a professorship at the University of Western Ontario, Canada, in 1972. The experimental petrology and associated analytical laboratories at Manchester (see Chapter 4) were very well endowed by the Natural Environment Research Council (NERC) and provided access to researchers from other universities as well as grants to Manchester to support postdoctoral workers. Some, like Drs A. C. Dunham, D. L. Hamilton and C. M. B. Henderson (who left a lectureship at St Andrews University to join the group) were initially supported on the NERC Experimental Petrology Grant for high pressure/temperature petrological research work before being appointed as Lecturers at Manchester.

Ansel Dunham was appointed to a lectureship in 1967 and promoted to Senior Lecturer in 1976. His earlier research included studies of the acid rocks of Rhum, the Skye lavas and the Whin Sill. Later he concentrated on industrial mineralogy, in which he was appointed to a new Chair at Hull in 1978 before moving to Leicester University. He retired in 1997 and sadly passed away in 1998. Dave Hamilton was an important member of the Experimental Petrology Group for almost forty years. He had great expertise in high pressure/temperature experiments and helped many colleagues and visiting researchers working in the 'basement labs'. Michael Henderson initially pursued research interests in thermal expansion and structural phase transitions in feldspars and feldspathoids as well as in petrological and geochemical studies of differentiated alkali basalt sills (rocks from the Shiant Isles and Arran). However, as noted in Chapter 4, he became a pioneer in the applications of synchrotron radiation techniques (notably synchrotron X-ray diffraction and X-ray absorption spectroscopy) to mineralogical and geochemical problems ranging from the structures of minerals and glasses, cation ordering in minerals, and solution complexes of metals in hydrothermal systems, to many other such topics. Over these years, promotion to Senior Lecturer, Reader and Professor followed and he was awarded a DSc by the University of Durham in 1992. His contributions were recognised by the award of the Schlumberger Medal of the Mineralogical Society, a society of which he also served as President. As noted in several sections of Chapter 4,

Michael also played key roles in the Manchester response to the Earth Science Review, and in the development of the Daresbury (synchrotron) Laboratory.

Other staff appointments during the '60s included Jack Treagus, a structural geologist with a particular research interest in the Dalradian rocks of Scotland, and Joan Watson, who maintained the traditions of the Geology Department in being appointed Lecturer in Palaeobotany (initially a joint post with the Botany Department). Her research has been mainly on important fossil Wealden flora studied using optical and scanning electron microscopy, and was recognised with the award of a DSc degree by Durham University in 1992. Pamela Champness was appointed in 1967 to develop electron microscopy in the department and was recruited whilst still completing her PhD at Cambridge. She made use of a newly installed electron microscope in the Geology Department, but also negotiated use of electron optical facilities in the Department of Materials Science. Her collaborative work with Dr Gordon Lorimer of that department on exsolution textures in feldspars and pyroxenes was very successful. They also got married in 1970! Treagus, Watson and Champness all spent their whole careers in Manchester, retiring as Readers (Watson and Champness) or as a Senior Lecturer (Treagus). Champness was not the only staff member to find marital bliss in the department; Joan Watson was already married to C. M. B. ('Michael') Henderson before they came to Manchester, and Jack Treagus married one of his former students, Sue Beech (latterly Sue Treagus), who went on to achieve distinction as a structural geologist and journal editor (Chief Editor of the *Journal of Structural*

*Geology*). Sue and Jack also collaborated on a book on the rocks of Anglesey (see Appendix 2).

This remarkably active period of 'comings and goings' also saw another two new academic posts. Dr Derrill Kerrick was the first person appointed in Manchester (1967) specifically to research and teach metamorphic petrology. Attracted to Manchester by the facilities for high temperature/pressure experiments, he moved on to a position at Pennsylvania State University in 1970. The second was Dr Eugen Stumpfl, a specialist in economic geology and ore mineralogy who had been at University College London (where he taught David Vaughan as an undergraduate). He moved on to universities in Hamburg and Leoben (Austria, his homeland) after a few years in Manchester (1966–1969). One final name to mention from the era of the '60s is Dr Peter Woodrow, a student of Jack Zussman who was appointed as Assistant Lecturer in Crystallography in 1966, but shortly after that moved to a post with the Geological Survey of Fiji. Whilst a student, he became a devoted member of the Baha'i faith, and his aim was to promote the faith abroad. This move enabled him to do so whilst also working in geology.

Until 1970, geophysics had been a largely neglected area of activity in the Manchester Geology Department despite, as noted above, the enormous contributions made in the Department of Physics associated with the work of Professor P. M. S. Blackett FRS. He was a pioneer in the development of palaeomagnetism, a subject essential for the plate tectonic revolution of the latter part of the 20th century. In the Geology Department, it was decided to 'invest' in geophysics through a new Chair in the subject, and Dr J. W. Elder of the Department of Applied Mathematics and Theoretical Physics, Cambridge University, was appointed. Elder had research interests in heat and fluid flow in the Earth, best summarised in his book *The Bowels of the Earth*. His studies of variable density flow simulators were associated with what became known as 'the Elder Problem'. Elder retired after thirteen years at Manchester and returned to New Zealand, his home country. A year after Elder was appointed, another appointment was made in geophysics but at Lecturer level, and to which Dr W. T. C. Sowerbutts was appointed. His research interests were very different to those of Elder, focusing on small-scale applied geophysics such as the use of ground-penetrating radar to locate buried mine workings. He was also interested in developing geophysical instruments such as magnetometers, and was a pioneer of computer-based learning in Earth sciences.

Also in 1970, A. B. Thompson, a Manchester BSc and PhD, was appointed as Lecturer to replace Derrill Kerrick, and was himself replaced by B. J. ('Bernie') Wood in 1973, on taking up a position as Assistant Professor at Harvard. Alan Thompson left Harvard some years later to take up a Chair at the Swiss Federal Institute of Technology, ETH, in Zurich where he has pursued a very successful career as an experimental and field petrologist. His work was recognised with the Bowen Award of the American Geophysical Union. Bernie Wood left Manchester in 1979, joining Rockwell International in the USA and then moving on to Northwestern University in Illinois, and then back to the UK to Bristol University, and latterly to Oxford, in a distinguished career as a mineralogist and geochemist recognised with numerous awards and election to the Royal Society.

The period from 1976 to 1984 saw modest growth during a period of consolidation. Dr A. E. Adams, a carbonate sedimentologist, was recruited from Oxford as a Lecturer. He was to spend the whole of his career in Manchester, and latterly his time was devoted to very ably managing the undergraduate teaching programmes. He was also involved (with Mac) in producing very successful 'atlases' of rocks and minerals seen under the petrographic microscope. Dr Richard Pattrick was appointed in 1978 as a Lecturer, replacing Ansel Dunham who moved to a newly created Chair in Industrial Mineralogy at the University of Hull. Richard Pattrick also spent the whole of his academic life in Manchester. He was recruited to promote teaching and research in economic geology and ore mineralogy, and later developed interests in spectroscopic studies of minerals, particularly those associated with the use of synchrotron radiation. Another aspect of his contribution (see Chapter 2) was in administration, serving as Head of Department and also Graduate Dean.

The '70s and early '80s were periods when funds available for making new appointments were very limited, so much so that the University Grants Committee (UGC) set aside funds for a programme of 'new blood' lectureships. These were awarded competitively and the Manchester Department was very successful in attracting these monies. Dr Giles Droop was appointed to a new blood lectureship in metamorphic petrology and Dr D. A. Polya to one in hydrothermal fluid geochemistry. As an outcome of the major reorganisation associated with the Earth Science

Review (see Chapter 4), three other 'new bloods' were transferred to Manchester from other universities: Drs David Manning, Andy Gize and Ian Lyon. Consequently, Manchester came to have the largest number by far of these sought-after posts.

The impact of these appointments was considerable and overlapped with the implementation of the results of the Earth Science Review discussed below. Giles Droop was appointed as a new blood Lecturer in 1982 and worked on the role of fluids in metamorphism and melting, laser probe dating of metamorphic rocks, and deformation and metamorphism in the Grampian (Scotland) Orogenic Belt.

David Polya, who joined the department from Australia as a Royal Commission for the Exhibition of 1851 scholar, completed his PhD in 1987 at Manchester on the Portuguese Panasqueira tin-tungsten deposit, initially working on the geochemistry of ore-forming hydrothermal systems before turning his attention to field, laboratory and theoretical studies in environmental geochemistry, particularly the challenge of arsenic contamination in shallow aquifers in Circum-Himalayan Asia. He also was appointed Associate Director of the Williamson Research Centre and Head of the Manchester Analytical Geochemistry Unit and, latterly, recognised by promotion to a Chair. He led multidisciplinary research/ research training partnerships with Europe, South-East Asia and India.

David Manning also completed his PhD at Manchester on an experimental petrology topic and, after several postdoctoral posts (in Manchester and in Nancy), landed a new blood lectureship at Newcastle to work on links between ore deposits and organic matter. He moved back to Manchester following the Earth Science Review, where he pursued a range of environmental projects from reactions between minerals in living systems in soils and plants, to deep geothermal energy in Northern England. He also returned to Newcastle some years later as Professor of Soil Science, and was awarded the Schlumberger Medal of the Mineralogical Society and elected President of the Geological Society in recognition of his work.

This period of modest growth was followed by the high drama of the implementation of the Earth Science Review (1988–1989), as described in detail in Chapter 4. The impact this had on staffing levels was unprecedented. The first changes were concerned with new appointments made in the year before the review's implementation so as to 'revitalise' the department. As described in more detail in Chapter 4, the retirement of Professor MacKenzie led to a search for a successor, and the upcoming retirement of Professor Zussman made possible a second ('proleptic') Chair appointment. David Vaughan was appointed to a Chair in Mineralogy and Charles Curtis to a Chair in Geochemistry as the result of a search. They then played leading roles in the review process, the result of which was an outstanding success for Manchester, with the academic staff numbers increasing by around 50% in a matter of months. This expansion involved the transfer of established staff from other universities, with isotope geochemists Professor Grenville Turner FRS and Dr Ian Lyon coming from Sheffield, Dr Ernie Rutter (a Reader in Structural Geology) and his wife Kate Brodie (Lecturer in Metamorphic Petrology) from Imperial College

London, and Dr D. A. C. Manning (new blood Lecturer in Geochemistry) from Newcastle University. In addition to the above (and transfer of Dr Gize from Southampton University), there were three new posts created: the specialist fields and appointees were geochemical spectroscopy (Dr Simon Redfern), experimental petrology (Dr John Clemens) and basin studies (Dr Rob Gawthorpe). Figure 9 shows the old (MacKenzie and Zussman) and new (Curtis, Turner and Vaughan) 'professoriat' at Manchester in 1989, immediately following the Earth Science Review.

Grenville Turner's move to Manchester, along with Ian Lyon and David Blagburn, was one of the most important developments for Manchester in response to the Earth Science Review. The group (much enlarged in the following years) uses isotope analyses of rocks and minerals, and the gases and fluids trapped within them (including extraterrestrial materials from the Moon, Mars, asteroids and comets), to understand the chemistry, physics and evolution of the Earth, the planets and other bodies from the solar system and interstellar sources. Grenville's contributions (not least in being the 'co-inventor' of the Ar-Ar isotope dating method) have been widely recognised nationally and internationally through election as an FRS, and numerous awards including the Royal Society Rumford Medal, the Leonard Medal of the Meteoritical Society, the Urey Medal of the European Association of Geochemists, and the Gold Medal of the Royal Astronomical Society. Ian Lyon had also been a new blood appointment, but in Physics at Sheffield before moving to Manchester, where he has been involved in developing new high-resolution isotope analysis methods and their applications to planetary science. Studies have ranged from the history of water on Mars, imaging pre-solar (SiC) grains, and the origins of terrestrial platinum deposits. Ian was promoted to a Chair in Cosmochemistry in 2011.

Kate Brodie also came to Manchester with the Earth Science Review alongside her husband, Ernie Rutter. Her main research interests are in the relationships between deformation and metamorphism studied both in the field and experimentally. As noted above, as well as people transferred to Manchester, there were the three new posts created at this time. The one in geochemical spectroscopy was initially filled by Simon Redfern who came from Cambridge, to where he returned in 1993 as a Lecturer and subsequently progressed to become Professor of Mineral Physics. (He also was awarded both the Max Hey and Schlumberger Medals of the Mineralogical Society.) A new post on the 'soft rock' side of geology was created in basin studies. The appointee, Rob Gawthorpe, eventually became Professor of Sedimentology and Tectonics, and Head of the Basin Studies and Petroleum Geoscience Research Group (2000–2010). His research was mainly in sedimentology, stratigraphy and tectonics of sedimentary basins involving landscape evolution, seismic and sequence stratigraphy, petroleum systems and reservoir modelling. He established and led a new MSc course in petroleum geoscience (1990). His third-year undergraduate course in sequence stratigraphy earned him a Partnership Award, and he also was awarded the Geological Society Lyell Fund as well as being the American Association of Petroleum Geologists Distinguished Lecturer for 2005. Rob left

Manchester in 2010 to go to the University of Bergen (Norway) as Professor of Petroleum Science.

Not surprisingly, this flurry of activity was followed by a period of consolidation. In 1992, Dr P. Selden, a palaeontologist on the staff of the Manchester Department of Extramural Studies working on fossil spiders, transferred to Geology, and in 1991 Dr Ray Burgess was awarded a Royal Society Research Fellowship (a post with a guaranteed tenure track academic position) to work in Professor Turner's Isotope Research Group. In a wide range of noble gas isotope projects, Ray has applied the Ar-Ar technique to mineral inclusions in diamonds to obtain genesis ages and to study the compositions of fluids trapped (as inclusions) in diamonds, and worked on argon diffusion in feldspars, and the chemistry of fluid inclusions in minerals from various terrestrial sources. In another area of investigation, he studied the noble gas geochemistry of lunar samples. His contributions were later recognised by promotion to Professor of Isotope Geochemistry.

In 1993–1994, changes were mainly associated with departures of established staff. Dr John Clemens left to a Chair at Kingston University (and in 2012 moved to become Head of Earth Sciences at Stellenbosch University, South Africa) and was replaced by Dr Alison Pawley. Alison conducts experiments on the stability of hydrous minerals at high pressures to understand, for example, the roles of $H_2O$ and $CO_2$ in subduction zone processes. Her contributions were recognised early in her career by the award of the Max Hey Medal of the Mineralogical Society of Great Britain and Ireland. Dr Roy Wogelius took on the position of Lecturer in Geochemical

Spectroscopy with the departure of Simon Redfern to a post in Cambridge. Roy has wide-ranging interests in aqueous geochemistry and environmental geochemistry, including applications to nuclear waste management and fossil pigment preservation, and a major interest in synchrotron-based techniques, as noted in Chapter 4. His more recent work has been undertaken both at the Diamond (UK) and Stanford University (California, USA) synchrotron facilities, the latter associated with the award of a Blaustein Visiting Professorship. Synchrotron work on fossil materials also led, in 2014, to establishment of the Interdisciplinary Centre for Ancient Life (ICAL) with Roy as its first Director. The mission of ICAL has been to foster collaborative work at Manchester University across all fields related to evolution.

One new appointment was that of Dr J. H. S. ('Joe') Macquaker, who had been a postdoctoral worker with Professor Curtis, and worked in sedimentary geochemistry and petroleum geoscience. A particular interest of Joe's was in the physical, chemical and biological processes influencing the formation of fine-grained sedimentary rocks. After a decade or so, he left Manchester for a post at Memorial University, Newfoundland, and latterly joined ExxonMobil as a Senior Researcher in Houston, Texas. He received the Wallace Pratt Award from the American Association of Petroleum Geologists (AAPG) in 2013 and was an AAPG Distinguished Lecturer in 2011.

A Royal Society Research Fellowship (leading to a tenured post) was awarded in 1994 to Dr Jamie Gilmour, another isotope geochemist and cosmochemist who had come to Manchester from Sheffield with Grenville Turner.

Jamie has an international reputation for developing techniques of resonance ionisation mass spectroscopy and applying these and other advanced analytical methods to problems in planetary science. He was given a tenured university staff appointment in 2002, and subsequently made Professor of Planetary Science. Jamie led the development of BSc and Masters courses in geology with planetary science, and latterly was Director of Teaching and Deputy Head of School.

In 1995, Dr Steve Boult, a Manchester graduate, was appointed Lecturer in Environmental Geochemistry, and Dr Sue Treagus became a Simon Industrial Fellow. Steve Boult has since played a key role as the Director of two important environmental science MSc degree courses (in pollution and environmental control, and in environmental sciences, policy and management) as well as conducting his own research in these areas. His other remarkable contribution has been in developing new instruments for monitoring water and gas quality in the environment and producing and marketing these devices through a spin-out company. The company went on to license them to international companies including Siemens. These water monitoring products are now used by all major UK water service providers; they have also produced the first battery-powered, in-borehole ground gas monitoring technology, enabling extensive and continuous site monitoring. These accomplishments were later recognised by NERC, who gave Steve their 2015 Award for Economic Impact. Another important initiative taken by the department at about this time involved appointment of Jonathan Redfern (from Oxford Brookes University) to a senior post (and latterly a

Chair) in petroleum geology. Jonathan was running a very successful MSc degree programme in petroleum geoscience, which he brought with him from Brookes. He also brought with him a North Africa Research Group, which addresses petroleum-related research problems from that region.

The period from 1996 to 2000 saw a number of appointments of Research Fellows taking up tenure track positions, including Steve Covey-Crump (Royal Society Fellow and then Lecturer in Rock Deformation Studies), Dave Hunt (University Research Fellow and later Lecturer in Basin Studies) and Michelle Warren (Colin Roscoe Research Fellow in Computational Molecular Environmental Science). Stuart Hardy was another Lecturer in Basin Studies, but he was wholly funded by industry (Saga Petroleum). Covey-Crump remained in Manchester, but the others moved on to jobs in industry or, in the case of Warren, associated with the Diamond synchrotron facility (see Chapter 4). This period also saw the appointment of Neil Mitchell, a geophysicist specialising in large-scale studies of the ocean floor, and the transfer of Dr John Nudds, a palaeontologist working at the Manchester Museum, into the Earth Sciences Department. His research activities have included work on fossil corals and dinosaur eggs. Dr Colin Hughes, who came from Sheffield with Professor Curtis as an Experimental Officer running electron microscopes, made the unusual transfer from a technical to an academic post, becoming Senior Lecturer and then Professor. Later on, Colin, who developed interests in environmental research and sustainability, took on major university roles concerned with the management of sustainability strategy.

Further strengthening of staff numbers took place in 2001 in the area of isotope geochemistry with the appointment of Chris Ballentine as a Senior Lecturer (and later, Professor). Chris had a very successful portfolio of research activities ranging from cosmochemical studies to environmental work, such as problems associated with the sequestration in former oil wells of 'waste' carbon dioxide from the combustion of fossil fuels. He was widely recognised as a leader in geochemical work by his election as President of the European Association of Geochemists (EAG). Chris left Manchester in 2013 to join the Earth Science Department of the University of Oxford. At the same time as Ballentine came to Manchester, another key appointment was made in association with the establishment of the Williamson Research Centre (WRC) for Molecular Environmental Science. Jon Lloyd, who trained as a microbiologist, was hired as a Senior Lecturer in Geomicrobiology and quickly expanded the WRC laboratories established for work concerned with mineral-microbe interactions in natural and industrial systems. Work in these areas soon involved colleagues specialising in mineralogy (Vaughan, Pattrick) and geochemistry (Polya, van Dongen), and Jon was soon promoted to Professor and became the second Director of the WRC following on from David Vaughan. The importance of his work has been recognised by the award of the Bigsby Medal of the Geological Society and the Schlumberger Medal of the Mineralogical Society. In 2014, Jon Lloyd was also placed by the Science Council in the top one hundred in rankings of UK practising scientists.

In 2002, Dr Merren Jones was appointed to a temporary post, particularly to help in the teaching of 'soft rock' geology.

She soon became a valued contributor to the department (school) and its teaching programmes and was granted tenure. Merren also helped to maintain another Manchester Geology tradition, that of married couples on the staff, when she married fellow Lecturer Steve Covey-Crump.

A final 'coming and going' in the period up to 2004 to mention involves Phillip Manning, who obtained his MSc degree in the department (in 1993) and PhD in Sheffield (1999) before returning to Manchester. His dinosaur research employing advanced imaging and scanning techniques has been at the forefront of work in the area, and Phillip has also been a great 'populariser' of the field of vertebrate palaeontology, with an international as well as national media presence. The novel imaging methods employed in his work have ranged from CT scanning and electron microscopy to techniques using synchrotron radiation (particularly involving collaboration with Roy Wogelius and with researchers in the USA). A good example has been the *National Geographic* Dinosaur Mummy Project, examining controls on dinosaur soft-tissue preservation. More recently, in recognition of his role as a populariser of science, Phillip Manning was appointed as STFC (Science and Technology Facilities Council) Science in Society Fellow, as well as Professor of Natural History at Manchester.

# CHAPTER 4

# THE RESEARCH SCENE

## THE EARLY YEARS

Geology began as an observational science, with rocks, minerals and fossils being described in the field or in 'hand specimens' taken back to the laboratory. Fieldwork has always been an essential component of geology degree courses and students have been taught how to record relationships between rock masses and present them as geological maps and cross-sections. Such work continues, not only in training, but in original investigations (see Figure 10, which shows members of a typical undergraduate fieldwork class). However, whereas the geologist in an area such as the British Isles working in the 18th and 19th centuries would have found much of Britain as yet unmapped, today the whole country has been mapped in a fair degree of detail. The early geologists in Manchester made significant contributions to this task, as will be evident from the accounts of staff activities in Chapters 2 and 3.

The work of mapping the geology of the British Isles was almost entirely conducted by the Geological Survey of Great Britain. Before taking the Chair at Manchester, William Boyd Dawkins was employed by them and was involved in mapping in Kent and in the Thames Valley. On the other hand, Pugh resigned from the Manchester Chair in order to become the Director of the Survey, and Holland was appointed Director of the Geological Survey of India in 1903. Manchester contributions in this area in the modern era have included work done by Jack Treagus and colleagues who were subcontracted by the Survey to remap an area in Scotland (the Schiehallion Map or 'Sheet 55W', remapped and published in 2000). Traditional geological mapping requires only the most basic of equipment used by the field geologist (compass-clinometer, hammer, hand lens and topographic map of the area being studied), who may work alone or with the support of a Field Assistant.

Although the earliest instruments enabling the geologist to study magnified images of minerals or fossils, the first 'microscopes', date back to the Middle Ages, a very important development in the laboratory study of geological materials took place in the 1850s. This was the fabrication of thin slices or 'sections' (of rocks in particular) which are translucent, and the ways in which polarised light could be employed to characterise and identify the minerals present in a sample. The discovery and development of these techniques is generally credited to the Sheffield-based geologist H. C. Sorby. However, Williamson contributed to the development and application of these methods, and another keen microscopist was Marie Stopes, shown at her microscope in Figure 8.

For the science of petrology, the single most important development has probably been the invention and refinement of the petrographic (or 'polarising') optical microscope. Not only could the identity of individual mineral grains in a thin section of a rock be determined by routine observations and measurements, but also their interrelationships; these could, in turn, provide information on the processes of formation or of subsequent deformation or alteration of that rock. Such petrographic studies could only be conducted on minerals that are translucent, whereas many metal oxide and sulphide (ore) minerals are opaque. A later development involved preparing a very flat, polished surface of an opaque mineral and observing (again using polarised light) the properties and interrelationships in reflected light. This reflected light (ore) microscopy was a particular interest of Manchester Professors Vincent, Pattrick and Vaughan. David Vincent used the technique in his classic studies of the opaque minerals in the famous Skaergaard Layered Intrusion in Greenland; Richard Pattrick and David Vaughan in numerous studies of ore deposits worldwide.

Until the middle of the 20th century, the chemical compositions of minerals and rocks could only be determined by using the traditional methods of 'wet' chemical analysis. These entailed dissolving a piece of the mineral or rock in question, using a strong acid, and then performing various procedures of quantitative analysis (such as titrations) to determine how much of a particular chemical element is present in a fixed amount of solution. As most rocks are made up of several minerals, before such analysis can be undertaken there is commonly a stage at which individual minerals are liberated by crushing, followed by separation based on density differences or magnetic properties. Obtaining good results entailed laborious and painstaking work. Professor David Vincent was a master of this type of analysis. One story concerning David was that, having agreed to meet with a junior colleague to instruct him in some of the techniques of wet chemical analysis, he was appalled when the colleague turned up wearing a smart new white laboratory coat to protect his clothes against any spillages. David, who was wearing no such 'protection', made it very clear that in his laboratory there would be no 'spillages requiring protection'.

We should also note that from the very beginnings of the science of geology, scientists have collected specimens for display in museums and for use in teaching and research. In particular, collections of fossils and minerals have been regarded as essential for teaching and also been very important research materials. Many have been collected by Manchester Geology staff over the years and an important contribution made by bequests from private collectors. A good example is the Harwood Mineral Collection, half of which was donated in 1975 from the estate of H. F. Harwood. Harwood had studied chemistry at Manchester, and was later a Reader in Analytical Chemistry at Imperial College London. He analysed many minerals in collaboration with geologists and featured as a co-author in many of their publications. The importance of the collections at Manchester was underlined at around the time of the Earth Science Review when Manchester was funded by the University Funding Council

(UFC, formerly the University Grants Committee or UGC) to be one of only five university 'Collection Centres' nationally given the task of safeguarding geological research collections. The Manchester Museum's Keeper of Geology, Dr Nudds, was appointed to head the centre. More recent donations have included the private mineral collection of former Manchester staff member Professor Bob Howie, who died in 2012.

## CONSOLIDATION IN THE EARLY AND MID 20TH CENTURY

The early to mid 20th century was a great period for research and discovery at Manchester University, although this was particularly in physics and associated with the names Rutherford, Bragg, Blackett, Wilson, Chadwick, Bohr and J. J. Thomson. Of these names, two in particular stand out for their relevance to geology.

W. L. Bragg determined how the diffraction of X-rays by crystalline materials could be used to determine crystal structure. The first materials he studied were minerals such as halite (NaCl) and pyrite ($FeS_2$). Bragg was appointed in 1915 and stayed in Manchester until 1937. Although Bragg is most often associated with Cambridge and the Cavendish Laboratory, he actually did much of his important work in Manchester (including determining many mineral crystal structures).

As discussed in Chapter 3, P. M. S. Blackett and his co-workers in the Manchester Physics Department made major contributions to understanding the Earth's magnetism which, in turn, led to acceptance of the theory of continental drift and thence to the revolutionary ideas of plate tectonics.

On a more modest scale, Manchester geologists played key roles in our understanding of what the eminent American petrologist N. L. Bowen called 'the evolution of the igneous rocks'. In particular, this included contributions to the classic studies of the Skaergaard Intrusion in Greenland and, later, to the petrology experiments set up by Mac and his co-workers (see below). Key people making up the Manchester Department in 1961 are shown in Figure 9.

## RESEARCH GROUPS, CENTRES AND INSTITUTES, AND THE RISE OF INTERDISCIPLINARY RESEARCH

Until the 1950s and the recovery of Britain after the Second World War, research at UK universities was still largely undertaken by the lone academic, sometimes with the support of a more junior colleague or the help of one or two postgraduate (PhD) research students. This was an appropriate scale of operation for work such as geological mapping, study of rocks, minerals and fossils under the microscope, or mineral and rock analysis in a chemical laboratory. However, it was insufficient in manpower to sustain the work becoming possible with the new technologies being used in experimental petrology and in advanced methods of microanalysis. These required groups of scientists with members ranging from Professors, Readers and Senior Lecturers, to Junior Lecturers, Postdoctoral Fellows and research students. A good example of one of the first such groups was that in experimental petrology at Manchester University.

## MAC AND THE EXPERIMENTAL PETROLOGY GROUP

As recounted above (Chapter 3), the late 1950s saw the appointment of W. S. MacKenzie (Mac) as a Research Fellow, but one who rose quickly to the rank of Professor. Mac used the expertise he had acquired working in the world's leading laboratory for experimental petrology (the Geophysical Laboratory at the Carnegie Institution, Washington, DC) to set up the first such laboratory in the UK in Manchester. Furnaces, pressure vessels and other such facilities were purchased or actually constructed in the new laboratories; these facilities enabled the high temperatures and pressures associated with the crystallisation or melting of rocks in the Earth's crust and upper mantle to be reproduced in the laboratory. The laboratories in Manchester required a large mechanical workshop, as well as other specialist technician support. Once established (and they were well supported by funds from the Natural Environment Research Council and its precursors), they attracted many excellent research students, postdoctoral scientists and eminent visiting scientists. One of the key aspects of the operation of the laboratories was in providing a service to the wider UK research community. Slightly later, facilities for experimental petrology were also established at Edinburgh University and work done in either of the labs was reported in an annual ('yearbook') publication as well as in major international journals. As regards the idea of a group rather than an individual scientist, experimental petrology at Manchester was arguably one of the earliest examples of such a group involving geologists.

## MICROANALYSIS: CENTRE AND SERVICE

Whether studying natural minerals and rocks or the products of laboratory experiments, there is a need to determine the chemical composition of mineral materials. An extremely important advance in mineral analysis occurred in the 1950s with the invention by the French scientist Castaing of the electron probe microanalyser (EPMA). In a 'microprobe', a beam of electrons (as in an electron microscope) is focused and directed so as to bombard a mineral grain in a polished section or polished thin section. This bombardment generates X-rays of an energy characteristic of the elements present in the mineral, and an intensity dependent on the concentration of that element. The electron beam only excites X-ray generation of a volume of sample of about a cubic micron, so can be used to obtain a quantitative chemical analysis of an individual mineral grain, or even a region within a grain. The Manchester Department of Geology was one of the first to have such an instrument, as well as employing specialist technical staff (latterly Fred Wilkinson, Tim Hopkins and Dave Plant). This investment was justified because Manchester, as well as catering for its own analytical needs, was funded to provide a service to the UK geological community. In addition, following investment in an instrument and joint acquisition with the Department of Materials Science, Manchester also became one of the centres for the applications of the electron microscope in geology. Access to both the transmission electron microscope (TEM) and scanning electron microscope (SEM) was also provided for Manchester and visiting scientists. The TEM could be used to acquire microstructural data at very high magnifications as well as information on atomic structures using electron diffraction, whereas the SEM could provide morphological data. Both microscopes could provide (semi-) quantitative chemical analyses of samples.

## MANCHESTER GEOLOGY AND SYNCHROTRON RADIATION: 'BIG SCIENCE'

The late 1950s and early '60s saw the development of techniques making use of the synchrotron radiation (SR) produced when electrons travel at near light speed in a storage ring that is tens of metres in diameter. As well as providing a polychromatic ('white') beam, this radiation can be 'tuned' to provide monochromatic beams ranging from high-energy X-rays to the infrared, and has the advantage of being an order of magnitude more intense than that from standard laboratory sources. The X-rays from the synchrotron can be used for diffraction experiments (XRD) and a range of spectroscopic studies, in particular X-ray absorption spectroscopy (XAS).

The first synchrotron facility built in the UK was at Daresbury near Runcorn, Cheshire; a very convenient location for access by Manchester University scientists. This was the first facility in the world to provide synchrotron X-rays. Although earlier work involved Manchester physicists and chemists, geologists soon became active in this area. Professor Michael Henderson, in particular, developed interests in XAS and XRD research on minerals and was later joined in the mid 1980s by Richard Pattrick and David Vaughan. A few years later, after the Earth Science Review, new appointments in the Geology Department included Alison Pawley, who worked on diffraction experiments involving hydrous minerals at the high pressures associated with the subduction of tectonic plates, and Roy Wogelius, who worked on the chemistry of mineral surfaces using state-of-the-art techniques including X-ray reflectivity measurements.

In more recent years, Michael Henderson became involved in the running of the Daresbury Laboratory through a joint appointment with Manchester, and was for a time the Director of those facilities. At the end of the 20th century the centre of UK synchrotron work moved to a new laboratory (called Diamond) in Oxfordshire, and Henderson served on several of the physical science working groups planning the new facilities. Synchrotron work by Manchester geologists continued (indeed, continues) at Diamond, and some, like Richard Pattrick, have continued to be involved in the administration of this kind of 'big science' through membership of key committees evaluating the many applications for time at the synchrotron, as well as chairing the main Science Advisory Committee.

## INTERNATIONAL COLLABORATION: THE EXAMPLE OF THE APOLLO LUNAR SAMPLES

The year 1969 witnessed one of the most outstanding of all human accomplishments: the first landing of a man on the Moon. NASA astronauts Armstrong, Aldrin and Collins and their successors brought back substantial amounts of 'moon rock' to study on Earth. Given the importance of these unique rock (and 'soil') samples, teams of scientists from around the world were asked to bid for an allocation of moon rock samples (and, in some cases, funding to cover the costs of investigations). One team of UK scientists led by Manchester Geology (with Jack Zussman as the Lead Investigator) was successful in a bid for samples. The 'Manchester Team' comprised J. C. Bailey, P. E. Champness, A. C. Dunham, J. Essen, W. S. Fyfe, W. S. MacKenzie, E. F. Stumpfl and J. Zussman. There were 140 allocations worldwide, of which fifteen were in the UK. Two were in Manchester; one to Geology and the other to Physics at UMIST to study luminescence emission spectra. The groups in the USA included one at the Massachusetts Institute of Technology (MIT), where future Manchester Geology Head David Vaughan worked on moon rocks using optical and Mössbauer spectra to study their chemistry.

The Apollo 11 samples retrieved from the Mare Tranquillitatis were handed to Principal Investigators at the Department of Scientific and Industrial Research (DSIR) headquarters in London, stressing the importance of security in transportation and when under laboratory investigation. Participants were to send results to NASA after three months, and subsequently present them for publication in the journal *Science* and at a conference in Houston, abiding by a 'gentlemen's agreement' (completely honoured) not to publish anywhere else in the meantime.

Soon after receipt, some lunar sample material was put on display under a microscope for the public for five days at the Manchester University Museum, and attracted long queues and a record attendance of about twenty-four thousand people. Unfortunately, all that was readily available at that early stage was what became known as 'moon dust' or lunar 'soil'; real lunar material, but not spectacular. The larger crystalline rock fragments studied later were remarkably fresh in appearance compared with most terrestrial specimens. Rocks known as breccias were also common, and among these and lunar soils were many spheres and fragments of glass.

Features of the main minerals determined in Manchester and elsewhere are consistent with the Mare Tranquillitatis being a very old, giant impact basin, subsequently filled by basaltic lava (3.6 to 3.8 billion years old), the near surface of which cooled rapidly, sealing in the lavas below, which cooled more slowly. Many later and lesser meteorite impacts left their craters and meteorite debris on the surface.

The Manchester group went on to work on samples collected in the Apollo 12, 14 and 15 missions, and results were published by the team members listed above and in papers co-authored by G. D. Nicholls (Manchester) and F. G. F. Gibb (Sheffield). The great public interest in 'moon rocks' led to numerous presentations by Manchester staff; Professor Zussman alone gave forty such invited talks. The Apollo programme was a spectacular example of international scientific collaboration, never seen before and unlikely ever to be seen again.

## THE UGC EARTH SCIENCE REVIEW AND ITS LEGACY

As outlined above, the University Grants Committee (UGC) Earth Science Review (1988–1989) was a revolutionary development for the subject in the UK; it was both unprecedented and surely never to be repeated. Although first steps of similar reviews in physics and chemistry were taken, they were never followed through. It was never stated publicly, but many assumed this was because the reorganisation involved, even for the relatively small Earth sciences community, was more costly than the UGC had expected.

Research was at the heart of the review, with a major goal being to bring together groups of scientists, along with the necessary funding to enable UK Earth science to compete effectively on the global stage. Departments were all invited to bid for support for reorganisations which could involve wholesale mergers or takeovers of existing departments, individual transfers at the request of a staff member, or expansions with hiring of new staff. The proposed new departments were ranked by a special UGC committee. Manchester's bid was ranked in the top group of six (with Cambridge, Oxford, Bristol, Edinburgh and Cardiff). As well as the resources needed to pay for various staff transfers, the review process also included the award (on a competitive basis) of monies for equipment. Manchester did very well in this competition, receiving £1.4 million for new research equipment, an amount exceeded only by Cambridge University; a grant was also made for building adaptations. The equipment included a scanning electron microscope, an X-ray photoelectron spectrometer (for mineral surface analysis), equipment for rock deformation studies, spectrometers for isotope studies, and a Mössbauer spectrometer to study the crystal chemistry of iron-bearing minerals. Separate grants from the Research Councils enabled the acquisition of a new electron microprobe and new equipment for X-ray powder diffraction.

The research groups established at the time of the Earth Science Review underpinned internationally ranked research work for the following decade (which also involved a major changeover of the professoriate – 'New profs for old!, as seen in Figure 11). Those groups and their Head Scientists were as follows:

- Physics and Chemistry of Minerals and Fluids:
  D. J. Vaughan
- Environmental Geochemistry and Geomicrobiology:
  C. D. Curtis
- Isotope Geochemistry and Cosmochemistry:
  G. Turner
- Basin and Stratigraphic Studies:
  R. Gawthorpe
- Structural and Petrological Geoscience:
  E. H. Rutter
- Palaeontology and Palaeobotany:
  P. Selden and J. Watson

## THE GEOSCIENCE RESEARCH INSTITUTE (GRI)

As part of the developments promised in the Earth Science Review, and to encourage collaboration between industry and the 'new' Geology Department, a Geoscience Research Institute was established in 1991. The GRI was set up to *promote and foster research activities (in particular interdisciplinary research) in geoscience at Manchester University and to promote collaborative links with industry and with public sector institutions.* Major spheres of activity were represented by five Institute groups, each under the direction of a group coordinator; these groups and coordinators were as follows:

- Petroleum Science:
     Professor C. D. Curtis
- Ore and Industrial Mineral Deposits:
     Professor D. J. Vaughan
- Mineral Sciences:
     Professor C. M. B. Henderson
- Environmental Geosciences:
     Dr A. P. Gize
- Isotope Geochemistry and Cosmochemistry:
     Professor G. Turner)

The Institute was under the overall control of a Board of Managers, chaired by David Vaughan and comprising the five coordinators along with senior staff from relevant departments of the university and UMIST. They were B. Beagley (Chemistry, UMIST), I. Douglas (Geography), C. D. Garner FRS (Chemistry), M. Hart FRS (Physics), J. Lee (Biology), G. Lorimer (Materials), and G. C. Wood FRS (Corrosion).

In addition, a very important role was played by a group of advisors to the board drawn from industry and public sector organisations:

- Dr J. Adams, Deputy Director of Research, English China Clays plc
- Dr P. Cook, Director, British Geological Survey
- Professor A. Leadbetter, Director, Science and Engineering Research Council (SERC) Daresbury Laboratory
- Dr D. M. Lloyd, Head, Physics Group, Coal Research Establishment, British Coal
- Dr B. McKinnon, Managing Director, VG Inorganics Division
- Dr C. J. Morrissey, Managing Director, Riofinex North Ltd
- Dr I. F. Price, Exploration and Production Manager, BP Research Centre
- Professor N. Richardson, Director, Surface Science Research Centre, Liverpool University
- Mr G. D. Spencely, Manager, Teesside Laboratories, British Steel
- Dr M. A. Sweeney, Head of Operating Company Liaison, Shell Coal International

The GRI provided a mechanism whereby Geology at Manchester University, and its research groups, could interact with key outside organisations in undertaking both applied research (for the petroleum, coal, industrial minerals, steel, and metalliferous mining industries) and more fundamental research (on mineral surface science, mass spectrometer development, applications of synchrotron radiation etc.). Critical to these developments was the synergistic relationship between the research groups set up in the wake of the Earth Science Review (see above) and those groups operating through the GRI. The prestige of the GRI was such that the official launch of the institute in May 1991 was performed by Sir John Cadogan FRS, the Research Director of British Petroleum (and later Director General of the Research Councils). With the launch of the Williamson Research Centre in 2001 (see below), the GRI was phased out to be replaced by the WRC and other centres.

## CENTRES BOTH RADIOCHEMICAL AND ENVIRONMENTAL

Manchester University was one of the very few organisations that maintained teaching and research activities of relevance to the nuclear industries in the latter part of the 20th century. A consequence of this was that Manchester, through its Chemistry Department, was able to bid successfully for funds from the nuclear industry in order to set up a centre for research in radiochemistry. This entailed the construction and equipping of 'hot' laboratories; it also led to strong relationships between Manchester and the nuclear industry, particularly British Nuclear Fuels Ltd (BNFL), who were the company that, in 1999, provided the £2.2 million needed to establish a Radiochemistry Centre of Excellence. Bids were led by the Manchester University Professor of Radiochemistry, Francis Livens. The involvement of Geology was through David Vaughan, who was co-principal investigator on the original bid. The funds were also used to support projects, including some jointly with Earth Sciences, concerned with the geochemistry and environmental chemistry of radionuclides and the safe containment of nuclear wastes. More recent decades have seen substantial research programmes in radiochemistry being developed at Manchester University (see Epilogue; Chapter 8).

Another important interdisciplinary initiative was the University of Manchester Environment Centre (UMEC), which was established in 1996–1997. UMEC was set up to act as a focus for the development of cross-departmental first-degree programmes and interdisciplinary environmental research and graduate training. The Geology (or Earth Sciences) Department led the way in the establishment of UMEC, with Professor C. D. Curtis as Director of Research and Graduate Activity. The university provided funding to give the new centre a physical home within the Earth Sciences (Williamson) Building. Adjacent space was also provided to house the Greater Manchester Geological Unit, which was funded by the relevant local government bodies and existed to provide geological expertise to local authorities. The UMEC umbrella structure existed to link research activities spanning Earth sciences, biosciences, geography, chemistry, engineering, mathematics, planning and landscape, and economics.

## THE WILLIAMSON RESEARCH CENTRE FOR MOLECULAR ENVIRONMENTAL SCIENCE (WRC)

The Williamson Research Centre (WRC) was established following a successful bid for UK Government funding of more than £3.3 million from the Joint Infrastructure

Fund (JIF). These funds were used to acquire new items of equipment and to establish new laboratories in areas including mineral surface science, geomicrobiology, environmental geoscience and advanced trace element analysis (see also Figure 16 for 'before and after' pictures of the refurbishment of the Chemistry Laboratories). The bid was led by David Vaughan, Professor of Mineralogy, who became the first Director of the centre. The WRC is an interdisciplinary venture involving Manchester University chemists and biologists as well as Earth scientists. The centre was officially opened in October 2001 by the newly appointed Chancellor of the University, the distinguished broadcaster Anna Ford (see Figure 17). Like the GRI, the WRC was under the overall control of a Board of Managers with representation from the collaborating departments (Professors Hillier and Roberts from Chemistry and Life Sciences, respectively) and members from different areas of geosciences including some appointed as Associate Directors (Jon Lloyd, David Polya and Roy Wogelius).

Molecular environmental science is the acquisition of knowledge at a molecular level of the materials (fluids, minerals, biomaterials etc.) at or near the Earth's surface, of the natural processes by which they are concentrated or dispersed, and of the effects on these processes of human activity (mining, pollution, waste disposal etc.). It is the fundamental science needed to underpin an understanding of our environment, and is essential in any attempt to predict the behaviour of environmental systems. It involves studies of natural fluids, minerals and their surfaces and how they interact with such fluids, and the interface between inorganic/organic and biological systems (bacteria, plants and higher organisms). The years since the launch of the WRC have seen substantial further investment in and development of the centre, notably in expanded laboratory facilities for research in geomicrobiology (bids led by Professor Jon Lloyd) and environmental aspects of radiochemistry (bids led by Professor Kath Morris).

## COSMOCHEMISTRY AND ISOTOPE GEOCHEMISTRY CENTRE

As part of the Earth Science Review reorganisation implemented in 1989, an isotope geochemistry group, led by Professor Grenville Turner FRS, moved from Sheffield University to Manchester Geology. At that time the other members of the group consisted of one Lecturer (Dr Ian Lyon), one Experimental Officer (Dave Blagburn) and several postgraduate students. They rapidly became established as one of the six research groups created in the 'new' Manchester Department, and expanded in personnel and equipment in the following decade. A major further expansion in their activities took place in 2002 with a successful bid for £2.1 million to a new Science Research Investment Fund (SRIF) set up by the Higher Education Funding Council for England (HEFCE) and as a successor to the JIF). This enabled construction of new laboratories and acquisition of new equipment which was reflected in the establishment of the University of Manchester Cosmochemistry and Isotope Geochemistry Centre, which was launched by Professor Sir Keith O'Nions FRS (Director General of the Research Councils) in April 2004 (see Figure 18).

# CHAPTER 5

# THE STUDENT (AND STAFF) 'EXPERIENCE' AND STUDENT NUMBERS

## ABOUT BUILDINGS, LABORATORIES AND CLASSROOMS

An important aspect of the student (and staff!) experience is the buildings provided by the university and the classrooms, offices and laboratories housed in those buildings. By the 1850s Manchester was a great industrial city, arguably the world's first and greatest. One of the city's wealthy merchants, John Owens, was persuaded to leave his fortune to found a college for young men. Owens College was to be like Oxford and Cambridge in what it taught, but would be non-sectarian and non-residential.

It was established in 1851 in the former home of Richard Cobden, a wealthy merchant and liberal parliamentarian. Here, space was provided for teaching geology; another important resource for teaching was the geological collections housed in the new (1835) museum in Peter Street.

A major advance in the development of the department took place about thirty years later with the Beyer Building (see Figure 12). The Beyer Building, together with the Manchester University Museum, was added to the main building of the university by the same architect, Alfred Waterhouse, and opened in 1887. It occupied the north side of what was to become the 'Old Quadrangle', to be used for teaching and research in botany, geology and zoology. Charles Beyer was born in Germany but moved to Manchester aged twenty-one and became a highly successful engineer specialising in building railway locomotives. He died in 1876 and his legacy generously provided for the building itself and the required appointments. A new building in the same style was added to the west side of Beyer in 1911 and housed the Botany Department.

The use of the Beyer Building changed immensely after the Departments of Geology, Zoology and, lastly, Botany moved (1959–1966) across Oxford Road to the newly constructed Williamson Building (see Figure 13). The stone structure remained, but subdivisions and amenities changed with time to serve new occupants and usages, with one major exception: the preservation of the Beyer Lecture Theatre (see Figure 14). This still has its seven rows of stout hardwood seating benches, with narrow steps for access at each end, and each row containing nine seats. The seats still have their old attached numbers 1 to 63. For some courses each student was allocated a number and a register of attendance was kept, and persistent absences noted. The Beyer Lecture Theatre is still in use. Jack Zussman recalls: *I do remember one student regularly bringing his dog to lectures and somehow getting its name into the attendance register.* Also on the ground floor there were research laboratories, while teaching laboratories and some staff offices were on upper floors, mostly specific to botany, geology or zoology, but occasionally shared. Some rooms today show where fireplaces were once present. Jack Zussman also recalls: *I shared a large room with MacKenzie [Mac] in the 1950s and on arrival each winter morning we would find a welcome coal fire burning, having been prepared and lit earlier by our cleaning lady [Mrs Blaymire].*

The basement was used mainly for palaeontology research and specimen storage. Present usages of the Beyer Building are for administrative purposes needed after the huge expansion of the university that has taken place. Examples are estate offices and meeting rooms, timetable organisation, car parking permits, environmental sustainability, and in the basement, the Professional Services Archives. Doors to the adjoining museum are, as before, generally locked.

The corridor walls used to house centrally controlled bells which rang very loudly at ten minutes before the hour for lectures to end and on the hour for next lectures to begin, allowing time for students to move between lectures. In some locations these could be unnecessary yet quite disturbing, and were known sometimes to become 'informally' disabled.

In front of the Beyer Building in the Old Quad, today there stands, appropriately at least for the former geology occupants, an extremely large boulder weighing more than twenty tons. This consists of an igneous rock called andesite, and was discovered during sewer construction in 1888 at a site described as the intersection of Ducie Street and Oxford Road, and was moved to the university. That was not its only journey. Such boulders are known as 'erratics' and have been carried long distances by glaciers, this one having come probably from Borrowdale in the Lake District, where andesite rock formations occur. The said location on Oxford Road is problematic, however, because Ducie Street ends long before the boulder would reach Oxford Road, even as shown on quite old maps.

Although we can only speculate about what life must have been like for a science Professor in the late-Victorian period, some insights can come from a permanent exhibit in the Buxton Museum, twenty miles or so away in the Derbyshire Peak District. Here, a room about the size of the original has been used to reproduce the study of Sir William Boyd Dawkins in his home in the Manchester suburb of Didsbury (see Figure 15). On display are numerous specimens and objects of interest along with papers and books. The connection with Buxton concerns the work done by Sir William and his assistant Wilfred Jackson, on materials found in nearby caves (particularly Poole's Cavern). The exhibit and the caves are well worth visiting and make for a pleasant day out in the hills of the High Peak.

The dramatic growth in teaching and research in science and engineering in the 20th century, and particularly after World War II, led to Manchester University constructing a large group of buildings to house activities in these disciplines on land opposite the original Victorian 'main' buildings, including the Beyer Building. The Victorian buildings still form the heart of the university and now house the offices of the President and other administrative functions. Geology moved into what was named the Williamson Building (see Figure 13) in 1958–1959.

## WILLIAMSON BUILDING

The Williamson Building, with its very striking sun sculpture, fronts onto Oxford Road but also has access (now limited) on its southern side along Brunswick Street. The University Master Plan for 2012 to 2020 is (at the time of writing) on its way to achievement, but the Williamson was built in the context of the preceding Master Plan. That plan included the movement of the major departments of the Faculty of Science and Engineering across Oxford Road to form a new Science Estate. When Geology moved in 1959, Electrical Engineering was already there on the opposite side of Brunswick Street in what was later to be named the Zochonis Building. The Simon Building (1962–1963, for Civil, Mechanical and Aeronautical Engineering), Roscoe Building (1964, with large lecture theatres, classrooms and offices), Chemistry, and Physics (Schuster) Buildings came later.

The Williamson Building was developed in four stages. Stage 1 has a basement, open to view through windows from outside across a (dry) 'moat', and five storeys above ground level; the lower levels were occupied by Geology and the upper storeys by the Mathematics Department.

Accommodation and amenities in Williamson were far superior to those in the Beyer, giving purpose-built laboratories for teaching and current research needs (chemical analysis, optical spectroscopy, X-ray diffraction, high-temperature/high-pressure experiments, rock storage, a mechanical workshop, etc.). Some attempt to associate the building with Professor Deer and his notable expeditions to Greenland was made by embedding some Skaergaard rock samples in the floor of the entrance hall. Unfortunately they were rather few and rather small for the purpose, and they disappeared when the entrance hall had a major refurbishment many years later.

Stage 2 of Williamson was a six-storey building with no 'moat', behind and parallel to the first stage, and with a main but controlled entrance from Brunswick Street. The third stage linked Stages 1 and 2, and not surprisingly became known as the Link, and the fourth and last phase of building (1966), extending Stage 1 towards Brunswick Street, was commonly referred to as the Gatehouse Extension, perhaps because of its position as seen on entering the Science Estate from Oxford Road. This would be in keeping with the 1950s Master Plan, which included the closure of Brunswick Street to traffic, a proposal that needed agreement from the Manchester city authorities, and was not to be obtained until very recently. The current 2012 plan has a new Brunswick Park at the centre of Brunswick Street, with work in progress at the time of writing (2017).

Geology has been in 'the Williamson' since it was first built. Other occupants arrived later, and some moved out for whatever reason. The first cohabitants were

Mathematics, who occupied the upper floors for several years but moved in 1970 to their new Tower Building, neighbouring the Williamson at that time. Their place was taken by Zoology and the major part of Botany, although a smaller part moved into the upper floors of Stage 2 and the Gatehouse Extension in 1966. Another small part of Botany and Zoology moved to some of the upper floors of Stages 1 and 4 in 1970.

Geology grew in numbers of students, staff and sources of finance, and expanded into the Link and Gatehouse. The basement, ground- and first floors of the Link part of the Williamson Building included a teaching laboratory, a small lecture room, stores, a mail and copying room, staffrooms and offices. The lower floors of the Gatehouse Extension gave the Geology Department much-needed space for more high-pressure and temperature experiments and advanced methods of chemical analysis, and for the large Harwood Mineral Collection, a small lecture theatre, and a library and students' reading room. Stage 2 provided space mainly for Zoology and Botany jointly, and Stage 4 for Botany on three upper floors. In the late 1980s the botanists and zoologists moved out of the building following a restructuring of their activities into a new Faculty of Life Sciences, and the University School of Law moved into the second, third and much of the fourth floors of the Williamson Building.

Substantially more space for Geology followed from the UGC Earth Science Review recommendations in 1988 (see Chapter 3). Space in the basement and at ground-floor level accommodated new laboratories for rock deformation, isotope geochemistry and cosmochemistry, and mineral

*Fig. 1*
*Portrait of*
*W. C. Williamson.*

*Fig. 2*
*Portrait of*
*Sir William Boyd Dawkins.*

*Fig 3.*
*Portrait of*
*Sir Thomas Holland.*

*Fig. 4*
*Portrait of*
*O. T. Jones.*

*Fig. 5*
*Portrait of*
*Sir William Pugh.*

*Fig. 6*
*Portrait of*
*W. A. Deer.*

*Fig. 7*
*Portrait of*
*E. A. Vincent.*

Fig. 8
*Marie Stopes examining thin sections
under the optical microscope in 1905.*

Fig. 9
*Manchester geologists in 1961.
The front row (left to right) are MacKenzie, Howie,
Morton, Deer, Simpson, Zussman, Isherwood and Paterson.*

*Fig. 10*
*Manchester undergraduate students examining*
*rocks in the field at Tyndrum, Scotland in 1980.*

*Fig. 11*
*New profs for old! Professors Turner, Curtis, Zussman, Vaughan*
*and MacKenzie in 1989 at the retirement of Jack Zussman.*

*Fig. 12*
*The Beyer Building,*
*Manchester University*
*in 2015.*

*Fig. 13*
*The Williamson Building, Manchester University in 2017.*

*Fig. 14*
*Beyer Lecture Theatre, Manchester University in 2017.*

*Fig. 15*
*Reconstruction of Boyd Dawkins' study in the Buxton Museum.*

Fig. 16

*The Chemistry Laboratories before (left) and after (right)
refurbishment as part of the Williamson Research Centre (WRC)
for Molecular Environmental Science.*

Fig. 17

*Opening of the Williamson Research Centre (WRC) by Anna Ford. Top:
Professor Vaughan talking with Anna Ford as the Vice Chancellor looks
on. Bottom: Professor Vaughan, Anna Ford, Peter Copley (grandson of W.
C. Williamson) and the Vice Chancellor (Professor Sir Martin Harris).*

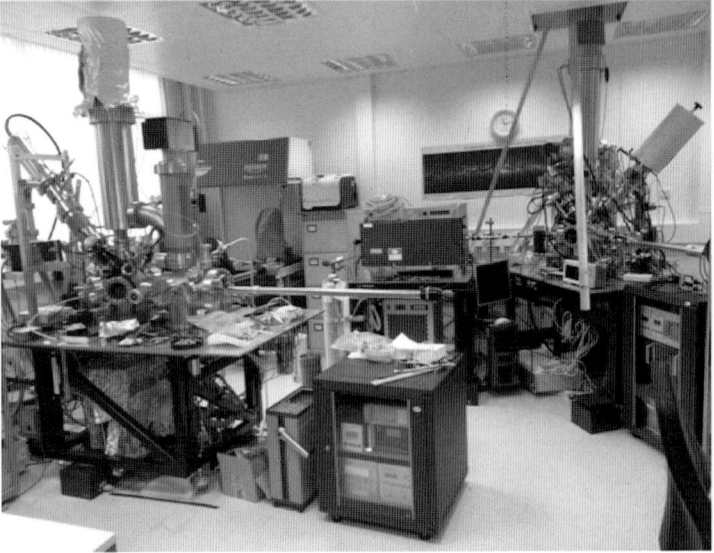

*Fig. 18*
*Mass spectrometers in the Cosmochemistry*
*and Isotope Geochemistry Centre.*

*Fig. 19*
*Atmospheric sciences research aircraft (BAe–146) operated on behalf of*
*NERC by the (Manchester) Centre for Atmospheric Sciences working*
*with the NERC National Centre for Atmospheric Sciences.*

sciences (X-ray photoelectron spectroscopy, Mössbauer spectroscopy, electron microprobe and scanning electron microscopy, X-ray diffraction and reflectivity). At first-floor level, laboratories for analysis of rock powders and solutions (XRF, ICP-MS, ion chromatography) were established. These facilities were further developed with the new funding for the Williamson Research Centre (2001) and included geomicrobiological laboratories. Research in sedimentology and basin studies was developed with computer imaging hardware, and isotope research further enhanced with the centre established in 2004 (see Chapter 3).

## UNDERGRADUATE STUDENT NUMBERS

In 1873, when geology was being taught as part of the curriculum for the BSc degree in natural history, twenty-one students attended classes in geology and four in mineralogy at the university. Evening courses were provided for the general public and were attended by thirty-three.

A full three-year honours degree in geology was established in 1881, but numbers registering were small and increased only slowly to about twenty distributed over first-, second- and third-year courses by 1930. These included a course predominantly on mining, given by G. H. Winstanley from 1905 and N. T. Williams from 1919 to 1939. Low numbers were not surprising, since geology was taught in very few schools, so was an unknown subject to many school-leavers. Numbers of honours geology students remained low through the 1930s and most of the 1940s, but ranged from sixteen to twenty-seven in their final (third) year during 1961 to 1970. From 1967 combined or joint Honours degree courses were offered in which students spent half of their time in each of two subjects. The most popular of these combined Geology with Physical Geography, but combinations with chemistry, physics, botany or zoology were also available. In 1990 a new honours BSc degree in geochemistry was introduced, and in 1992 another on the environment and resources. Total numbers of final-year students for geology and combined honours degrees had reached fifty-eight, and the total for all three years 150 by 1984, and two hundred by 1990, partly reflecting the overall expansion in university places throughout the UK and changes in Manchester's resources and staff numbers. Table 1 shows undergraduate numbers averaged over ten-year periods for each single or combined honours degree course. Thus, for example, in any one year between 1880 and 1889, there was an average of twenty students and the total number in that decade would have been 10 x 20 = 200. Adding these total numbers for each of the decades gives the grand total of students for the period 1880 to 2003, which is about five thousand single honours and two thousand combined honours students.

Notable rises in undergraduate student numbers in the 1960s and 1970s were associated with the beginning of combined honours geology and geography, and other combined honours courses and degrees; also there was overall increased funding for university expansion. The late 1980s and early 1990s saw a major rise as a result of the UGC Earth Science Review, and also the addition of new undergraduate courses and degrees. For example, in 2002 a new taught course for a four-year undergraduate degree of

MEarthSci was introduced, and many more were added, but beyond the designated time range of the present historical account.

In addition to the above courses for students specialising in geology or 'combined' geology, for many years, there was available a separate lecture course in geology for first-year students as a subsidiary to other main or combined subjects. This was a popular choice, the numbers attending increasing gradually over the years from about twenty to over two hundred per year. In the early 1920s a course on mining and metallurgy (including economic geology) was attended by twenty-five students from a variety of disciplines, and a little later the Department of Geology began providing a first-year course in geology especially for students of civil engineering from both Manchester University and the College of Technology (later to become the University of Manchester Institute of Science and Technology, or UMIST). Provision of this course had been facilitated by the part-time appointment in 1923 of Edgar Morton, a well-known consultant in applied geology (see Chapter 3). He was later assisted, and after retirement replaced, by C. W. Isherwood. The number of students registered was twelve in 1934, and by 1951 it was 110 from Civil Engineering, plus thirty-one from Town and Country Planning, and twelve from Building and Surveying. In 1960 the total registered was 212.

At the time of writing (2017), the School of Earth and Environmental Sciences is offering BSc degree programmes in geology, geochemistry, geography and geology with a year abroad, environmental and resource geology, geography and geology, geology with planetary science, environmental science, environmental science with a year in industry, and environmental science with a year abroad. Also offered are MEarthSci degrees (which are four-year undergraduate qualifications) in Earth sciences and in geology with planetary science, and both BEng and MEng courses in petroleum engineering. An interest in geology, particularly amongst the undergraduate students, was also fostered by the establishment of a Geology Students Society as early as 1905. The society organises field excursions and social activities for its members.

## POSTGRADUATE RESEARCH STUDENTS

The first postgraduate students to carry out supervised research on geological projects in Manchester did so in the years 1906 to 1913 towards the degree of MSc, and later, PhD. Manchester was unusual amongst large university Geology Departments outside Oxford and Cambridge in not offering any taught-course MSc degrees before the 1990s. The situation changed with the advent of a petroleum geoscience MSc brought to Manchester by Professor Jonathan Redfern in 1995 (see below), and two new environmental science MSc degrees initiated at about that time. At the time of writing (2017), MSc courses are being offered in petroleum geoscience, pollution and environmental control, environmental sciences, policy and management, and applications in environmental sciences.

With the growth in staff numbers following developments such as the Earth Science Review, there was a rapid growth in the 1990s of PhD students. This was also helped by the availability of greater numbers of studentships to cover the fees and maintenance of doctoral students. Some of these came from industry, but most from the Research Councils. Although most came from the Natural Environment Research Council (NERC), the Manchester Department was unusual in attracting support from the Particle Physics and Astronomy Research Council (PPARC) for work on extraterrestrial materials, and from the Engineering and Physical Sciences Research Council (EPSRC) for work on materials and at large facilities such as synchrotrons. The other source of PhD candidates and funding for them was from abroad.

Postgraduate student numbers developed approximately as follows:

- **1950:** 3
- **1960:** 16
- **1970:** 31
- **1980:** 26
- **1990:** 56
- **2000:** 70
- **2003:** 78
- [**2010:** 101; **2017:** 251]

These numbers are total postgraduate students at all stages of their training for the years specified; from 1990 the numbers include students attending new MSc taught courses.

## TEACHING QUALITY AND ITS 'ASSESSMENT'

For those who were undergraduate students through the '60s, '70s and '80s, and certainly before that time, there was little or no monitoring or assessment of the quality of university teaching. Indeed, there was often little or no instruction provided to newly appointed staff on how to lecture or organise laboratory or field classes. That started to change in the '70s with 'induction courses' for new staff; often these were quite modest, lasting for just a few days. This situation changed quite dramatically in the '90s when the Higher Education Funding Council for England (HEFCE), responsible for allocating Government monies to run the universities, began to take an active interest in the quality of the teaching that it was funding. In particular, a Teaching Quality Assessment exercise was introduced by HEFCE which all institutions funded by them had to undertake. This, of course, included geology at Manchester.

The Teaching Quality Assessment (TQA) was a major exercise for those being assessed. A team of about six 'experts', made up of academics from other universities and some from industry (a number of such teams were appointed, each to assess several institutions), were put in place by HEFCE. They visited departments being assessed, spending a full week during which time they sat in on lectures, laboratory classes and tutorials, interviewed students to get their views on the courses on offer, and studied large amounts of paperwork describing courses, examinations, dissertations and other forms of student output. At the end of this process, a report was produced which went to the Vice Chancellor (VC) of the university concerned and to the relevant Head of Department. There was also a 'grade' given alongside the report. Specifically, a department was rated 'Outstanding', 'Satisfactory' or 'Unsatisfactory'.

The Manchester Geology experience of the TQA might best be described as traumatic. The best institutions were almost all rated as 'Outstanding', but Manchester Geology was rated only 'Satisfactory'. The report, when received, contained incorrect statements and was generally very negative. The then-Head of Department, David Vaughan, with the support of the VC, decided to appeal against this outcome. As a result Manchester went through the whole process again, with an entirely new team spending a week in the department. The result was that Geology at Manchester was indeed appropriately rated 'Outstanding'.

# CHAPTER 6

## SOME PERSONAL RECOLLECTIONS FROM FORMER STUDENTS AND STAFF

We invited a number of former staff and students each to write a short contribution recalling some aspects of their own experience of geology at Manchester University. The choice of subject matter was left entirely to them and their contributions have not been edited or revised in any way. They are published below as received.

## JACK ZUSSMAN

*Assistant Lecturer, Lecturer and Senior Lecturer, Manchester University, 1952–1962. Reader in Mineralogy, Oxford University, 1962–1967. Professor and Head of Department of*

*Geology, Manchester, 1967–1989. Professor Emeritus 1989 – present.*

My recollections go back to 1952 when I joined the then small Department of Geology as Assistant Lecturer, bringing academic staff to seven. Honours Geology and research student numbers were correspondingly low, so interaction between staff and students was easy, but benefitted also from contact hours in laboratory work as well as lectures, and still further from courses gaining experience of rocks, minerals and fossils in the field. Each course involved students and staff being away together, usually for one week, giving more opportunities for getting to know each other.

The importance of field work was appreciated from the early days of the Department when occasional 'Field Lectures' were given by Professor Boyd Dawkins, and the number of such courses attended by each student increased over the decades, involving the Easter as well as Summer vacations, and adding a course in the field devoted to geological mapping. Accounts of alumnae in later life about their university experience frequently recount memorable stories from their field courses. Such stories and other recollections emerge on occasional visits to the Department from individual ex-students, and on group reunions such as that of the graduation year 1969, held at Chester in 2015, to which I was pleased to be invited together with Professor Henderson, a more recently retired colleague. Their reunion this year is to be in Iceland, organised by two Icelandic ex-students, but also very interesting geologically.

The graduation year of 1972 met in Manchester in the summer of 2017 and visited a number of research laboratories in the Williamson Building. After lunch they were led by John Pollard on a part of the Simpson and Broadhurst tour of 'Building Stones in the City of Manchester'. The following day some were led by the reunion's organiser Phil Portus on a field trip in Derbyshire.

## Facilities and Collegiality

There probably were, even in the early fifties, Departments (though not Geology) that had a room or area in which staff members could have a coffee or tea break, either individually or at a recognised time to meet briefly, but I do remember a few of us in Geology going across Burlington Street to the University Staff 'Senior Common Room' at such times. There we would sometimes join the 'Top Brass' (Registrar, Bursar and University Librarian) in conversation. Lunch with a varied menu, with waitress service and at a reasonable price, was available in a large Staff dining room. Although there was no official seating scheme there was, more-or-less by custom, a 'top table'.

In the 1960s, after the Department's move across Oxford Road to the Williamson Building, there was space to designate a 'tea room', with volunteers organising supplies and payments. This was well used and valuable as an informal meeting place for staff and research students, and continued in one location or another in the Williamson Building for many years. The University Senior Common Room, however, with its dining room, and lounge with newspapers and current journals, served collegiality on a large scale in attracting staff from many university

Departments, but it was not continued in later years when the building was demolished and the site put to other use.

## Departmental Social Events

No doubt there have always been 'social events' devised and operated by members of the Department of Geology but it is worth noting how their nature changed with time.

A popular summer event, and perhaps least changing, was the annual Staff against Students cricket match played at the Hough End recreation grounds. My own level of cricket prowess was strictly limited so I did not make it to the Staff team. I was however allowed, when I was Head of Department, to keep the score, using a very professional score book. As far as I remember, in the 1950s to '80s the Staff team won most of the matches, but not by too big a margin.

A regular winter event in the 1950s and '60s near to Christmas was the Hot Pot Supper which took place at the university's sports pavilion in Wythenshawe and is well remembered as a very robust occasion. The main feature was a rather wild game of indoor rugby, but also some impressive competitive feats involving the climbing enthusiasts and some items of furniture. The Head of Department was expected to give a speech followed by downing a pint of beer. The Hot Pot Supper was open to all staff and students, but female students, knowing its reputation, rarely attended. At the other extreme is today's mode of celebration which is dinner and dancing at one of the best hotels in Manchester, popular with, and often organised by, female staff and students. Over the years between, there have been: Christmas meals at one or other of Rusholme's curry restaurants (attended by about 80); a party, sometimes themed and fancy dress, at a University site; a turkey buffet meal at one of the Department teaching laboratories. At other times of the year: A Staff summer barbeque on the lawn around the Williamson Building; a party for 3rd year students on finishing exams, and Staff at the Zussman home (in years when numbers involved were still manageable); refreshments at the Department for Geology for graduates and their guests on degree days.

I recall from my first several years at Manchester some staff social events of a kind that could and did occur with the small numbers involved. There was also once a year a lively evening party for academic staff at the home of Alex and Margaret Deer in Bramhall. Professor Deer would also occasionally decide on an afternoon that it would be nice to go together to a restaurant in Alderley Edge for a meal that evening. Hasty phone calls to home resulted in most couples managing the short notice, and car arrangements were made. I do not recall the food but have no difficulty about the wine because it was bought and ordered by Alex Deer and we always exchanged smiles because we knew that after some contemplation it was always going to be… Nuits Saint George! One winter Wednesday afternoon (designated for student sporting activities; no lectures) he suggested we went together to see Manchester City play at the nearby Maine Road Ground. About one Department it was said that its Head insisted that all his Staff supported 'City'!

## Geology Ancillary Staff

In early days of the Geology Department numbers of staff and students were very few, and there was little or no equipment so that there were no technical or office assistants. When I joined the Department in 1952 however, there was a Secretary to the Head of Department, who dealt with his correspondence, phone calls, etc., and covered his typing needs including papers for submission to publishers. She would also type papers for other academic staff when time was available.

There was also one Laboratory Assistant (or 'Lab Steward'). At that time, it was Walter Paterson, previously a laboratory assistant at Rossall School in Fleetwood (founded 1884), of rather shabby appearance but memorable character. He was always available to prepare laboratories before each use and to help lecturers and demonstrators in any way needed, despite having suffered a bad injury to one arm. He was getting on in years but seemed to be there at all times of the day and well into night, acting it seemed as unofficial 'watchman' to the Beyer Building and beyond. There had been a spate of thefts of lead from the University's roofs; he spotted one thief because of the way he was carrying a bucket. Paterson somehow continued for several years beyond retiring age.

Having mentioned an early secretary, and first Lab Steward, the first workshop technician was Mr Marcovitz, previously employed as a welder. Items he constructed were good but tended to be rather heavy. Professor Deer also had a laboratory assistant to help with his chemical analyses, and there was a technician, appropriately named Cutner, who cut thin sections and made polished sections for transmitted and reflected light microscope work.

## ROBERT IXER

*Honours Geology BSc, PhD, Manchester, 1965–1972. Lecturer at Leicester, Aston and Birmingham Universities. Later an independent scientist working in geoarchaeology including important contributions to understanding Stonehenge through petrographic studies.*

I was in the department from 1965 until 1972 (indeed graduated as part of the 1969 year that likes reunions) during a time of academic transition, scientific revolution and excitement. In retrospect it was the very end-times for the avuncular/paternalistic but safe university, that had little changed in ethos from the end of the Second World War and was to be startled by the student revolutions of the very early '70s. A university serviced by a single main frame computer and a 'well-equipped department' beginning to use the microprobe and mass spectrometer routinely but where wet-chemistry remained important.

Although the department was not much touched by university politics it was by the great 20th-century geological revolution. We must have been the last undergraduate year where continental drift was briefly discussed, land bridges were invoked and the origin of andesites was inexplicable – my first recollection of plate tectonics (indeed hearing Benioff for the first time) was a talk by Bill Fyfe a couple of weeks before finals. I clearly remember his excitement (and exactly where I was sitting) as he explained what it meant (when he asked if there were questions, we said will it be in the finals exam!). He was a pervasive influence, for undergraduates – he taught a course called 'I guess this must be mineralogy' – but more so for his research students and the staff, in his early stable isotope work, low-grade metamorphism and his proselytising that 'water' was the single most important aspect of geology, oh, and his dancing at the Christmas parties.

But in addition there was the prestige of the department investigating moon rock (with its accompanying security), the presence of experimental pressure vessels in the basement, Percy the plesiosaur in the foyer and the availability of a well-rounded mixture of geological experts just along the corridors and the pervasive sense of Whiggish movement.

It was a wonderful time to be involved in geology and especially in Manchester geology; it remains, perhaps my best, certainly most favourite, time and I never think of those years but I feel warm and grateful for the teaching, tolerance and kindness, not least by being given the freedom to become a petrographer, to be allowed to dabble in small shiny minerals and archaeology. With experience my skills have improved over the last 50 years but the Manchester geology staff taught me the soul of our science by their example. I shall always be indebted to them and being there at that time.

## DAME ANGELA STRANK

*Honours BSc, Geology, University of Manchester, 1972–1975. PhD, Manchester, 1976–1980. DSc, Manchester, 2017. Chief Scientist and Head of Downstream Technology, BP Oil Company; also a Board Governor at the University of Manchester.*

I went to Manchester to do a degree in Chemistry but soon changed my major to Geology after being inspired and 'hooked' on the subject by the charismatic Dr Fred Broadhurst, Senior Lecturer in Palaeontology, who took me on my first field trip to Castleton in the Peak District looking for fossils.

I started Hons Geology at Manchester in 1972 – just as the North Sea Oil and Gas industry was opening up and the major Forties field had been discovered.

It was very difficult for women to become professional geologists in those days, especially in the oil and gas industry; the world of equal opportunities was in its infancy. Furthermore, it was illegal for women to work offshore in the UK (until this changed later in the '80s), but I was determined to be a geologist and work for either BP or Shell, or the British Geological Survey – two of which I managed during my career.

There were only 2 females in my Honours Geology group at Manchester and about 40 men, but it didn't make any difference – we all got on really well. I had two special student friends, Brian Thompson and Paul Jenkins, both of whom I am still in contact with from time to time. Paul became a lecturer at Oxford Brookes University and Brian spent his career in the coal industry in South Wales, where he is to this day. Paul, Brian and I did all our practicals and field mapping together and were inseparable for much of the time. At the end of second year we hired a cottage in Snowdonia with another student, Mike, and spent 6 weeks living together as we mapped the challenging peaks of Snowdonia in continuous pouring rain! It wasn't until about 20 years later that I climbed to the top of Snowdon again and actually saw, for the first time, the view on a clear day.

I absolutely loved being in the Geology Department. It was such an inclusive, welcoming and friendly department with some great lecturers and some real characters! I was encouraged to reach my career goals all through my degree and the PhD that followed, which gave me the confidence I needed to follow my aspirations. Fred Broadhurst was my brilliant PhD supervisor (and inspirational ex-Bevin Boy), and he encouraged me every step of the way. Fred had another PhD student at the time – a lively character called Dave Mundy who was a fun-loving postgrad who demonstrated in our geology practicals. He became a good friend, and we ended up working together for some years at the Geological Survey in Leeds and for BP in Canada for a while. As well as Fred, I was encouraged enormously by Mike Henderson, who still looks the same today as he did 35 years ago! Mike always took a keen interest in his students, and amongst all the geochemistry I learned from him, he taught me not to be knocked back by any disappointments, to stay confident and to 'move on', in his typical pragmatic, straight-talking and clear-thinking style.

The lab technicians, Ian Nicholson and Harry Locke, were the most helpful, amusing team – they couldn't do enough for us. Ian, in particular, was a star when I wanted to borrow numerous samples of fossils or minerals to teach my extramural geology evening classes in various parts of Greater Manchester. He was also a great companion in the field driving the Land Rover and helping collect enormous quantities of limestone all over Northern England to turn into thin sections for me – usually in torrential rain I recall! Sue, the photographer, had endless patience with all my photos of microscopic foraminifera – there were thousands of them, and between us we developed them all by hand and spent endless days in the darkroom in the basement. Pam Collins and Cyril Guildford were a strong team who managed supplies and stores prudently and brilliantly to make sure every penny was well spent by the students. Dear Prof McKenzie gave me my first job as a Research Assistant in high-pressure and temperature mineralogy in the basement – where we made something akin to emeralds I recall, and his PA Pat Crook became a good friend as she helped me type my rather long PhD thesis for 25 pence a page and with lots of Tippex! Professor Zussman managed a thriving department and made Manchester famous through the iconic DH&Z – which was the book I consulted most (by a long way) during my 7 years at Manchester. Prof Zussman's office was run like a smooth machine by Joan and also Elaine (who eventually married Harry Locke, mentioned above).

I shared my PhD lab with Helen Fisher, Pat Cossey and Kevin Mason. I am still in touch with Helen and she visits from Australia regularly. We all meet at Garsington Opera in the summer with Professor Mike Henderson, Joan Watson and Judith (the department librarian in the '70s) for an annual reunion. Kevin left to be a museum curator in his native Wales and Pat went to Keele University as a lecturer I believe.

There were, of course, some student 'high jinks' but my friends were a fairly studious crowd so very little comes to mind. I do recall though on my 21st Birthday a few of the chaps in my year insisted we went to the union bar for a drink at lunchtime. I only went to the union bar a few times in my whole time at Manchester, and on this occasion somebody 'spiked' my half-pint of lager and lime (a trendy drink in the day) with a vodka, and the rest of the afternoon's lectures were somewhat of a giggly blur despite Robin Nicholson's best efforts to engage me in structural geology! He wasn't amused, needless to say... On another occasion on a field trip to Boscastle with Bernie Wood and Mike Anketell some of the students had an impromptu party in the Youth Hostel after 'lights out' and that didn't go down too well with the proprietors, needless to say. Luckily I wasn't part of it so managed to avoid the public dressing-down and humiliation at breakfast the next morning when the students involved were asked to leave the hostel immediately!

Field trips are like indelible ink on my memory – the first one to the Isle of Arran was breathtaking for me – a rather naïve young woman who had not travelled north of Cambridge before coming to Manchester to study. Glen Rosa and Lochranza were inspiring, unlike the severe chest infection I acquired that week due to the obligatory and incessant pouring rain, intensely cold Easter weather and lack of spare money to buy any field boots or effective waterproofs – my plimsolls didn't survive the field trip and were left ceremoniously somewhere on Arran. Before the second-year field trip I splashed out on a pair of John Brown leather field boots, which I still have and wear to this day, and I have hiked and walked in them all over this country and various other parts of the world.

Recently, I was lucky enough to meet Sir David Attenborough in London at a reception. We talked about how we both loved Earth Sciences but both somewhat 'struggled' at university with crystallography, crystal optics, Miller Indices and the like... These seemed to be subjects that students in our Manchester class either 'got' or they didn't. One day Jack Zussman was explaining some complexities from Deer, Howie & Zussman and suddenly the mysteries of crystallography became clear to me, something clicked into place and I never looked back. Jack Zussman and DH&Z have stayed with me throughout my career – moving all over the world with me as I have lived and worked overseas, and I still refer to this extraordinary encyclopaedia of knowledge from time to time, although my microscope days are long over.

Other lecturers, Ansel Dunham (with his pipe), John Wadsworth, John Essen, Morven Simpson, Robin Nicholson (who always wore a very smart grey suit, sometimes embellished with blackboard chalk dust), Bernie Wood (who introduced a more casual and contemporary dress code to the department), Joan Watson (my female role model who looked like Emma Peel from *The Avengers*), Pam Champness, Bill Sowerbutts (famous for teaching us to find culverts outside the Williamson building, and to whom I am grateful for preparing me for a lifetime of looking at colourful, complicated seismic sections!), Tony Adams, Richard Pattrick – the list is endless and too many to name here. I owe them all a great deal. Without them, their inspiration and guidance I wouldn't have been able to achieve all I have done in my career. This is why I am honoured to now be a member of the Board of Governors at the University of Manchester so that I can give something back to an institution to which I owe so much.

Manchester as a city was a very different place in those days. Whitworth Hall was cleaned of all its industrial black grime and soot the year I started and it looked like a brand-new Victorian building – so splendid and grand as a centrepiece of the university. In the '70s we lived as students through the 3-day week, the disturbing miners' strikes of 1974, frequent power cuts, studying by candlelight, petrol rationing and the oil crisis, the binmen's strike – with rubbish piled high in the streets everywhere, the Winter of Discontent, the building of the 'yet to become famous' Arndale Centre in the City Centre, and the decline in economic wealth in Manchester. My rent was £9 a week, a gallon of petrol was 50p, a bottle of wine also 50p, and my total daily food allowance 30p including a 5p cup of hot chocolate and a Kit Kat for lunch from the machine in the Williamson Building foyer – which is still there, I see!

I could go on for pages more but I'll stop here as this gives you a flavour. Thank you for inviting me to send you my brief recollections. My student days at Manchester were very happy indeed.

## JOHN POLLARD

*Honours BSc, Geology, Oxford University; PhD, University of Durham. Lecturer and Senior Lecturer in Palaeontology, University of Manchester, 1962–2001.*

I joined the Department in October 1962 after a first degree at Oxford and a PhD at Durham on Coal Measures microfossils. This was I believe David Vincent's first appointment to strengthen palaeontology after the retirement of Sidney

Straw in 1958 and followed Alex Deer's last appointments of Robin Nicholson and John Dewey in 1960.

In my second academic year 1963/64 I had full responsibility for palaeontology teaching in all years totalling 45 hours while Fred Broadhurst was on sabbatical leave at the Palaeontological Museum in Oslo. This involved 1st-year Palaeontology, 2nd-year Invertebrate Palaeontology, 3rd-year courses Environmental Palaeontology and Stratigraphical Palaeontology. After the examinations in July 1964, together with Robin Nicholson and Dr John Stanley (Postdoctoral research assistant in Palaeontology 1963–67?), we took the 2nd-year Honours and Combined Honours to Oslo for two weeks. This was the first foreign field trip from the Department and was led partly by Fred and staff of the Oslo Palaeontology Museum. It was a memorable trip with group accommodation (bunks in the forecastle!) on the Fred Olsen ferry *Braemar* and youth hostel accommodation in a converted warehouse in Oslo! We not only studied the Lower Palaeozoic stratigraphy of the Oslo region but also many classic igneous rock types described and named by Brøgger (1890–1916).

The next year, 1965, another two-week excursion of 2nd and 3rd years together in the Assynt region initiated detailed field mapping training on rocks of the Moine Thrust Zone. This training continued annually between 1st and 2nd years for Honours Geology, led by myself and Mike Anketell from 1965–68 at Cross Fell in Cumbria. Later this course moved for many years to the Coniston area of the Lake District where I last assisted at it in 1998.

Amusing recollections from this period include the student who removed a farmer's stone walling hammer from Cross Fell believing it to be a 'lost' geological hammer. Three months later we were visited by the police who returned it to the farmer – declared value 5 shillings! On a field trip to Skye a couple of vegetarian and animal-loving students 'rescued' two orphan lambs wandering by the roadside and brought them back to Manchester to graze on the lawns of Hulme Hall. When it was pointed out to them that they had committed sheep rustling, the lambs were presented to Pets Corner in Platt Fields Park – with no questions asked!

During the 1960s increase in specialisms of new staff broadened the courses within the Honours BSc degree. This included Sedimentology (Mike Anketell, 1965), Marine Geology (Geoff Nichols), Structural Geology (Robin Nicholson and Jack Treagus, 1966), Geochemistry (John Esson, 1964), Petrology (John Wadsworth, 1964), Geophysics (Bill Sowerbutts, 1966, and John Elder, 1970), Palaeobotany (Joan Watson, 1966) and diverse hard rock aspects from members of MacKenzie's HP/HT research group, Mike Henderson, Ansel Dunham, Bill Brown, Bill Fyfe, David Hamilton, Eugen Stumpfl and others.

By the mid 1970s the staff decided on the restructuring of the degree courses to permit a core and choice of options scheme, especially in the 3rd year. Many new special options were offered from 1976 including personally Trace Fossil and Environments (1976–1999), the first such course in a British university. Other specialisms included Metamorphism (Giles Droop,1976), Carbonate Sedimentology (Tony Adams, 1975?) and Ore Mineralogy (Richard Pattrick, 1980).

In the late 1970s and early 1980s foreign field work was again introduced with trips to Mallorca and the Boulogne-Rouen area of Northern France. The University also entered into a five-year 'timeshare' agreement for accommodation on the Isle of Mull, utilised by both Honours and Combined Honours field classes.

The next major reorganisation and incoming of staff was the Earth Sciences Review of 1988. Geochemistry was strengthened by Charles Curtis, Mineralogy by David Vaughan and Structural Geology by Ernie Rutter and Kate Brodie. Further changes into the 1990s resulted from the retirements of Morven Simpson (1983), Geoff Nichols and Fred Broadhurst (1990) and Jack Zussman in 1989. Personally I inherited stratigraphy teaching and Faculty Tutorship for Combined Honours Geography, Chemistry and Biology (1983–1990) from Morven Simpson.

The highlight of my career was hosting the International Trace Fossil Workshop in the Department in July 1999 when we were able to demonstrate our trace fossil collections to 35 international trace fossil workers. Also with the help and sponsorship from former research student Andy Taylor and his company 'Ichron', we took this international group for field work on trace fossils in the Jurassic rocks of the Yorkshire coast.

## JEFFREY FAWCETT

*Honours Geology BSc, 1954–1957; PhD, Manchester University, 1958–1961. McRae-Quantec Professor of Geological Sciences, University of Toronto, Canada, 1964–2002, and Emeritus Professor.*

Arrived Manchester Sept. 1954 with a background in geology from 2 years of Grammar School Ordinary Level and 2 years Advanced Level geology taught by a man who had done a few geology courses at King's College, Newcastle.

Incoming 1st-year geology students were invited to a field trip run by 3rd-years and possibly research students. We went to Robin Hood's Bay area on the Yorkshire coast; this provided an excellent introduction to classmates.

Classmates – Tony Ashton, Dave Hamilton, Tom Bolton, Keith Jeffries, Malcolm Farghar (sp.?), Geoff Potts, Norman Wardlaw and one I only remember as 'the Rev'.

Found my way to lodgings on Monton Street in Moss Side. It was only later that I became aware of the shady reputation of Moss Lane East.

Located the Professor's office in the Beyer Building and had my introduction to Prof W. A. Deer who advised on course selection for honours geology. (? Geology, Chemistry, Zoology, Crystallography.) The dark corridor outside the office was cluttered with sledges and camping gear from recent field work (Baffin Is. or Greenland?).

1st-year geology (Prof Deer) was largely a repeat of my Advanced Level courses at school – patterned on the first edition of the Arthur Holmes text.

Chemistry was OK but Zoology was a mystery; I had no previous biology course. Prof H. Graham Cannon usually arrived at the classroom door fresh from a pint or two at the College Arms (correct name?). I was not well equipped for crystallography.

1st-year field trip – Arran led by the Prof, who had not been there before!

2nd year – Palaeontology course from Dr S. H. Straw, an expert in Silurian stratigraphy; specialised in blackboard drawings of life-sized graptolites.

Mineralogy was taught mainly by the Prof from handwritten notes for the much rumoured 'BOOK'.

Prof Deer broke his arm (hand cranking the car in his garage?) but still taught mineral optics and tried to draw a 3D version of the optical indicatrix on the blackboard with his left hand.

Stratigraphy Morven (sp.?) Simpson – very easy going.

Sedimentary petrology taught by Dr Howie – mainly how to identify heavy minerals in grain mounts.

Igneous Petrology – taught by Dr Nicholls – went like the wind and lectured largely from memory but very clear and understandable.

Metamorphic petrology (Bob Howie?).

Third year – a new man on the scene – W. S. MacKenzie came from the Geophysical Lab, Washington, DC. Taught a senior seminar course in experimental petrology for specialists; very talkative and approachable.

Howie told us of the death of N. L. Bowen who featured greatly in the MacKenzie seminars.

The Beyer Building was a rabbit warren of staircases, corners, landings and impossible-to-find offices – those of Bob Howie and Geoff Nicholls were up a steep flight of stairs behind the blackboard of a second-floor laboratory (quite a challenge for Bob).

Mac started to assemble a hydrothermal laboratory in the basement – not far from the machine shop; the machinist was known to me only as 'Mac workshop' – a Polish(?) man who had limited command of English and did drawings for his projects by scratching with a chisel on his workbench.

I liked the hands-on aspect of the lab and stayed in the evening to cut panels for temperature controllers and pressure gauges.

Rock storage was along past the new hydrothermal lab – floppy cardboard boxes were very dusty; crushing for rock analysis was also carried out in that area.

Thin sections were made by the imperturbable Mr Henderson – very obliging and could work very quickly and carefully when he wanted.

No recollection of that era could omit reference to Walter Paterson – Departmental Steward, possibly the only such position in the whole University.

Very strong, despite missing one hand (lost in a lab explosion at a private school when mixing acids?). He had keys to most of the University doors – a legacy of his days as a fire warden during WW II. He enjoyed night-time visits to the Medical School labs where the 1st-years did their human dissections; he had the only telephone outside the staff offices but guarded it carefully. He brought down the house when trying to mend an electrical fixture in a lab while standing on a chair – sparks flew everywhere and Paty's false teeth kept popping in and out of his mouth. He was a kind man who could be a great help to those he liked; I was fortunate to be in that group.

Prof's assistant – did classical chemical analyses in a small room adjacent to Prof's office.

2nd year field trip – Welsh borderland/Church Stretton – led by Fred Broadhurst.

No U/G classes on Wednesday afternoon – students encouraged to get involved in sports. Geology football team played most weeks in winter.

3rd-year field trip – led by Prof Deer to Ardnamurchan; MacKenzie also came along.

3rd-year Mapping course – led by Fred Broadhurst – 2 weeks to map areas near Weymouth/Lulworth Cove.

Classmate Dave Hamilton married long-time girlfriend Margaret; for the honeymoon, she went to the USA and Dave and I took a student flight to Munich and hitched for 3 weeks in Austria and Switzerland. On one stretch of the road we passed 'Mac' (sorry, I don't recall his full name), a suave undergrad in the Dept, perhaps a year or so behind me. Dave saw him just in time to lean out of the window of the truck we were riding and offer Mac a hearty obscenity.

Somehow I got a first-class BSc and was awarded a DSIR scholarship. Prof Deer offered a research project on Isle of Mull. Not much guidance – 'Go and collect some of the Tertiary lavas and see what you can make of them.' Registered for MSc, later transferred to PhD.

Prof – 'You had better analyse those lavas.' Bob Howie gave instruction on classical chemical analysis in a small ground-floor lab; the lab benches had to be cleared completely every Friday so Mrs Blairmeyer (sp.?) could dust and polish the wood surfaces. Bob suggested I look into rapid (colorimetric) methods – and I set up some rapid methods (based on USGS reports and the work of E. L. P. Mercy in London, but results were inconsistent).

Did some mineral separation (magnetic and heavy liquids) and a lot of 2Vs with guidance from Ron Croasdale.

Got involved in the Rag Day with a float we built in the Dept basement and also in Men's Student Union (through Dave Hirst and Brian Rushton?) and ended as treasurer in 1957–59 (approx.). I met medical student Sylvia Manton (later Women's Union President) who became, and still is, my wife. I also was talked into being treasurer of a Student Union sponsored Arts Festival promoted by Geoff Furness.

Back home (Blyth, Northumberland), a local Rotary Club rep (John Johnson) asked if I was interested in applying for a Rotary Foundation scholarship. It came through and, mainly with advice from MacKenzie and Zussman, I gave Penn State as my preferred destination. I worked for a year under E. F. Osborn on the effect of controlled oxygen fugacity on basalt crystallisation.

Moved to new building (Williamson), where Neville Bradshaw improved my rapid analytical methods.

Got engaged to Sylvia and she finished Medical School during my year at PSU.

Returned to Manchester and finished PhD degree in 1961.

New equipment had arrived, including the XRF – I did the first calibrations for Fe determination with glass bead starting materials – with a lot of help from Jack Zussman.

Met research student John Rucklidge who later joined me as a colleague at U. of Toronto.

I thought I had it made when Prof Deer awarded me the Dept Research Assistantship – 100 pounds/yr. I went to the Lloyds bank on Oxford St. and asked for a loan as I was about to get married. I was hoping for few hundred pounds. The manager was very sympathetic and shouted to the teller, 'Write him up for 100 to be repaid in full when he gets his first pay cheque!'

Overall recollection – a friendly atmosphere with helpful, very knowledgeable faculty and fellow students. The education and guidance given to me was a wonderful introduction to academic geology that helped me obtain a PDF at the Carnegie Institution Geophysical Laboratory and a career-long faculty appointment at the University of Toronto.

Prof Deer was appointed to the Min Pet Chair at Cambridge during my last year (1961). In the summer of that year he took an overnight train to Manchester for my final oral. On arrival he greeted me at the Department door: 'Where the… is your thesis?' It was upstairs on his desk and the defence was postponed until late afternoon.

Finally, a lasting memory is of Sylvia and I attending the 1961 May Ball in Whitworth Hall as guests of Jack and Judy Zussman during our last few months in Manchester.

PS: Who was that dapper man who appeared in the Dept on a regular basis after parking his Rolls-Royce in the quad?

## JANE SPOONER (NÉE GALL)

*Honours BSc Geology, Manchester University, 1969–1972. MSc Environmental Resources, Salford University, 1972–1973. Co-Chair, M.Plan International Ltd, 2015 – present.*

My godfather, Harry M. Fairhurst, was instrumental in my decision to read geology at Manchester. He was the architect for the Williamson Building, which was opened in 1964, and arranged for me to visit the department. My visit turned out to be an informal interview. Professor MacKenzie showed me around and introduced me to Fred Broadhurst who, at that moment, was working inside the glass case where the fossil Plesiosaur, affectionally and universally known as Percy, was installed. Fred explained that Percy was suffering from incipient pyrite disease and was being anointed with Savlon cream. I do not remember much more about my visit but none of the universities where I had formal interviews could match my introduction to the geology department at Manchester.

My godfather designed several other science buildings at the University and thought very carefully about the function of his buildings and the well-being of the people who worked in them. During the early planning stages for the Williamson Building, he must have had many meetings with Professor Deer, then head of the department. On one of those occasions, he was amused to see that Professor Deer, absentmindedly, had his shirt on inside out. Professor Deer was, of course, one of the authors of 'D. H. and Z.', *An Introduction to the Rock-Forming Minerals* by Deer, Howie and Zussman, which was our constant reference as undergraduates and has been mine periodically throughout my career. Harry also considered carefully the materials for his buildings: the Williamson Building, with the Schuster, Chemistry and Simon Buildings, are referred to as the Science Quadrangle and share extensive use of copper cladding and roofing (now refurbished with a replica copper cladding system). He commissioned the sculptor, Lynn Chadwick, to produce the *Manchester Sun* which is installed above the Oxford Road entrance to the

Williamson Building, and he would have been responsible for choosing, and probably commissioning, the sculpture by Michael Piper on the roof of the lecture theatre of the Schuster Building and the *Alchemist's Elements* mosaic by Hans Tisdall in the lobby of the Chemistry Building. I am sure he knew that the functionality of the laboratories and lecture halls of this group of buildings needed to be relieved by artwork that was integral to the design. Harry was not above a degree of playfulness, however, since elsewhere, he designed one of the first buildings to have a tree planted on its roof.

It was a particular pleasure to revisit the Williamson Building in September 2017, when a group of fellow graduates had an informal reunion. New benches replace the original teak or mahogany in the laboratories, but other features remain in place and in use, like the built-in storage and display cabinets which line the west side of the corridors of the Earth Sciences wing, and which Harry and Professor Deer must have agreed upon together.

## WARD CHESWORTH

*Honours BSc and MSc Geology, Manchester University, 1955–1960. PhD, McMaster University, 1961–1964. Research Associate, Penn State University, 1964–1967. Professor of Geology, Guelph University, Canada. Professor Emeritus, Guelph University, 2002 – present.*

*[Note that this contribution draws directly upon 'amusing material' written by Ward Chesworth for his light-hearted column in* Earth *Magazine, published by the American Geological Institute.]*

The Past is a Foreign Country
Ward Chesworth

The words above, often misquoted as 'the past is another country', are L. P. Hartley's, and the saying is literally true for an emigrant. For me the past is 1950s Manchester, a Gradgrind-grim chunk of the Industrial Revolution, as often as not shrouded in a fourth state of matter called smog. The city was still emerging from wartime austerity into the beginnings of a post-war boom. Rationing wasn't completely over until 1956 by which time Harold Macmillan was assuring us that we'd 'never had it so good'. Towards the end of the decade, a combination of the death of industry, the Clean Air Act and the sandblaster was beginning to reveal the true colours of buildings that had been uniformly blackened by 150 years of coal smoke.

I came to the University's Department of Geology tempted by a hike in the Grampian Highlands inspired by Robert Louis Stevenson. *Kidnapped* was a favourite book of my teenage years and the episode called the 'Flight through the Heather' tells how 17-year-old David Balfour and his renegade companion Alan Breck Stewart, are pursued over the Pass of Glencoe and across Rannoch Moor, with King George's Redcoats in murderous pursuit. Two hundred years later and a year older than the fictional Davie, I decided to hike this route, though in the opposite direction. This still has the reputation of being the only true wilderness in Britain and my experience of the landscape and its geology is as close as I've been to an epiphany. It was here that the last ice sheet of the Quaternary glaciations wasted away, and isostatic rebound still lifts the region at a rate of about 10cm per century. So that's why it's the very highlands of the Highlands, with Ben Nevis at 1,345m just north of the Moor. Had I known the Plate Tectonic narrative I would have been even more bowled over, but that was at least 20 years into the future. Totally unsuspected in the fifties was the fact that the volcanic rocks I saw exposed in Glen Coe were evidence of a dramatic collision 400 million years earlier, when a collection of continental bits and pieces called Avalonia was jammed beneath Laurasia. Five major eruptions evacuated some 1,000–3,000km$^3$ of material from an underground magma chamber and the resulting collapse structure would probably have rivalled Yellowstone Caldera in Wyoming.

The hike through the heather wasn't all joy and enlightenment however. Although I wasn't pursued by Redcoats, I was chased and caught by numerous cloudbursts. A regular soaking guaranteed that I spent many miserable hours soaked to my undergutchies. I almost felt like the character in the novel *John Splendid*, who said that if he had the choice of staying for an extended time on Rannoch Moor or of being hung from the gallows, he would tell the hangman to go fetch his rope. Even so, I became a 'fresher' in Geology at Manchester in the autumn of 1955 when the Department was still on the west side of the Gothic quad opposite the Williamson Building, its home two years later. The three-headed monster Deerhowieandzussman was yet to be written, and though introductory geology was taught by Professor W. A. Deer himself and R. A. Howie introduced us to rocks and minerals, Jack Zussman was kept under wraps until the second year.

Palaeontology was the province of S. H. Straw, then on the brink of retirement, and though all three of our earliest lecturers were good, humane teachers, they were rather sedate and somewhat lacking in the enthusiasm of youth. But in the New Year, Fred Broadhurst burst into our lives as Straw's replacement. Full of the fire and brimstone of a newly minted PhD and not much older than ourselves, he was a Bevin Boy who had served his national service 'down th'pit', and had obtained a doctorate working on non-marine lamellibranchs from the Coal Measures he worked amongst. He was never Dr Broadhurst to my generation, however, but simply Fred – not a mark of disrespect, but one of real affection for an inspiring man. I owe it to Fred for convincing me to stick with geology when I was about to quit.

At the end of the decade, I obtained a student visa at the Canadian High Commission in Liverpool and left Manchester for McMaster University in Hamilton, Ontario. I remember being told by one of the officials in Liverpool that I would really feel at home in the smog of Hamilton. That was a calumny on the clean city I found there, and when I asked what clothes I should wear when I arrived in the midwinter cold, he lied again. 'More or less what you're wearing now,' he said. When I got off the Canadian Pacific liner *Empress of Britain* onto the frigid docks of St. John, New Brunswick, the emasculation of brass monkeys came readily to mind. But I survived that scrotum-tightening winter, and 50 subsequent ones in North America. Now in my dotage I write 'Geologic Columns' for *Earth Magazine*, published by the American Geological Institute. For their Manchester-related content, I'll finish with two of these.

## Lumpers, Splitters and
## a Scot with Red Whiskers

Names are important. Take Rumpelstiltskin for example. He's a dwarf in one of the Brothers Grimm fairy tales who would lose his magical abilities if his name were ever disclosed. For the price of her firstborn child, he helps a girl to fool a king into thinking that she can spin straw into gold and thereby solve his majesty's cash-flow problems. Rumpelstiltskin does the spinning and the grateful king makes the girl his queen. Demanding his reward, the dwarf reappears when the queen gives birth. She pleads for mercy, the dwarf relents and gives her three days to guess his name or she loses the child. If he'd had a regular name like Jim, Fred or even Gunter, the task would have been hard enough, but his name is way out at the extreme end of the spectrum of all monikers. Yet, by subterfuge she discovers the answer in time, and the dwarf storms off, madder than Yosemite Sam each time a cigar explodes in his face.

In science, taxonomists have the job of managing names. They take the objects that we study, and construct classifications wherein each distinct type of object is assigned to its own pigeonhole and tagged there with a specific name. It may be a familiar name like limestone, or something barbarous like rumpelstiltskinite. (Sadly, that doesn't exist, but out on the edge of the spectrum similarly monstrous mouthfuls lurk, Jacupirangite for instance.) The aim is to give some semblance of order to the chaos of nature, and when it's done right a classification is an important route to new knowledge.

Mendeleev showed this by using his Periodic Classification to predict elements unknown to the chemists of his day.

Controversy arises in deciding how many names are enough? Take the International Giraffe Working Group for example. No, it isn't an old *Monty Python* skit where tall animals gather in a committee room with a high ceiling. It's a serious group of specialists trying to decide whether to leave giraffes safely lumped together as one species under one name, or to split them into six species with six names. 'Lumpers and splitters' is the informal taxonomy of taxonomists first suggested to Charles Darwin by Hewett Watson in 1855. A couple of years later Darwin passed the terms on to J. D. Hooker in a letter and explained that *those who make many species are the 'splitters', and those who make few are the 'lumpers'.*

Excessive splitting could lead to problems, though. If a power-mad taxonomist kept on splitting hairs ad absurdum, he'd wind up with a classification that resembled an advanced case of logorrhoea, the kind of thing guaranteed to drive a working geologist nuts. C. B. Hunt staged his own rebellion against this tendency when he considered the plethora of names invented for minor igneous intrusions. He expressed his displeasure by sarcastically concocting one more, cactolith, which he described as a quasi-horizontal chronolith composed of anastomosing ductoliths whose distal ends curl like a harpolith, thin like a sphenolith, or bulge discordantly like an atmolith or ethmolith. He insinuated it into his 1953 USGS Professional Paper on the Henry Mountains of Utah, and from there it crept under the radar into the first edition of AGI's *Glossary of Geology*. Unfortunately, some humourless jobsworth banned it from all subsequent editions.

Here's where William Scott MacKenzie comes in. He left the Geophysical Laboratory in Washington, DC for the University of Manchester in the UK. To me and my fellow undergrads, the association with the 'Gee-Whiz' Lab only added to the Scottish charisma that Mac undoubtedly had in abundance. Whenever I visited England later in life, I would drop in to see him for a good gossip and one time I found him in the smoky gloom of his dusty office, peering down a petrographic microscope working on photo-micrographs of thin sections for one of his atlases. He invited me to name the volcanic rock he was examining, and seeing plagioclase and pyroxene, I pronounced it basalt. 'A fat lot you know, Chesworrrth,' he bellowed, with a fine Hibernian rolling of the r. 'You hav'nae lairnt a damn thing since you left here. Can you no see it's a mugearite, man?' Mac was a splitter – to call it a basalt labelled me a frivolous and imprecise lumper. Later, over lunch with the inevitable pint of warm beer, he added, 'Och, you're just a bluidy arm-waver.' But he was a humane man; he let me keep my firstborn.

As time went by, Mac's red beard turned to grey and a friendship spiced with forty years' worth of gratuitous insults ended when he died in 2001.

(Adapted from *The Rumpelstiltskin Factor*, *Geology* magazine, May 2015.)

### Fred, Percy and the Second Jurassic Coast

For me there were two notable events in 1847. One was literary, the other scientific. The literary event was the publication of the first Mills and Boon Romance, over a century before they actually existed. It set the formula in which a shy, self-effacing girl goes to work for a domineering, untamed hunk of a man, falls in love with him, and after a convenient house-fire rids him of the mad wife he kept stashed in the attic, he's eventually tamed by love and circumstance and the story ends in matrimony. *Reader, I married him* announces the heroine finally in Charlotte Brontë's *Jane Eyre*.

The notable scientific event was the death of Mary Anning, the first great female 'fossilist' – the term used before 'palaeontologist' became common. Mary's life of 47 years was spent in and around Lyme Regis, at the foot of the cliffs of Jurassic rock that mark the coast of the English Channel there. Being female, poor and a religious dissenter, a formal education was denied her. She and her brother learned a little from their father, a carpenter who supplemented the family income by collecting and selling fossils. Mary was clearly the more academically inclined of the kids and went on to teach herself enough geology and anatomy to become a palaeontologist in all but name. She was only twelve when she painstakingly excavated a 'crocodile' that her brother had found, and which was later named the world's first ichthyosaur ('fish-lizard' in palaeo-speak).

Just out of her teens she discovered the partial skeleton of another world first, another marine reptile but more reptilian than the ichthyosaur. So 'near-lizard' (i.e. plesiosaur) became its name. Later she excavated other plesiosaurs including the one that defines the genus. Louis Agassiz of Harvard's Museum of Comparative Zoology named two fossil fishes after Mary – a rare contemporary recognition of her contributions to science. She was finally called a geologist in a biography published 78 years after her death from breast cancer. In 2010 a panel of the Royal Society of London named her one of the 10 most influential women in British science. Largely because of her and the dinosaurs she discovered, her part of England is now called the Jurassic Coast as a kind of 'come hither' to tourists. It helps of course that Michael Crichton made the Jurassic the sexiest period of the stratigraphic column.

The cliffs around Lyme Regis where Mary collected are at the southern end of a band of Jurassic rocks that can be followed diagonally across England for almost 500 km forming the Cotswold Hills along the way to their final destination on a second Jurassic Coast where the North Sea confronts Yorkshire. For my generation of alumni of Manchester University's Department of Geology, this second Jurassic Coast is associated with a charismatic teacher of palaeontology called Fred Broadhurst. Fred was a 'Bevin Boy', who chose as a teenager to do his national service working in a coal mine. When he went on to become a palaeontologist he wrote his doctoral thesis on the non-marine lamellibranchs he found there. I was arguably Fred's least attentive student, being more interested in a female botanist in the class, but his enthusiastic, colourful lecturing style carried me through to a passing grade. When I became an academic myself, I discovered that I had learned a great deal from Fred's example, and consciously adopted his teaching technique as my own.

In 1960 Fred took his undergrad class to the second Jurassic Coast. Below the beetling cliffs of Robin Hood's Bay near Whitby, one of the students hammered off a knob of rock and brought it to be identified. Fred and his assistant Jas Potts agreed that it was vertebrate bone and later confirmed that a large Jurassic reptile lay buried in the tidal zone. When Fred went back to Manchester, he obtained funding to return to the coast to excavate the fossil and bring it to his lab. I was a geochemistry graduate student by then, but managed to insinuate myself into his party of six volunteer labourers. For two days from dawn to dusk, we worked between tides to excavate blocks of the Jurassic, backpacked and dragged them to the cliff top and then headed back to Manchester. After cleaning up and trimming away unwanted rock, the fourteen-foot fossil was re-assembled like a jigsaw puzzle. After 20 years in a glass case in the Williamson Building at the University, 'Percy the Plesiosaur' now rests in Manchester Museum just across the road.

In case you`re wondering what happened to the botanist who diverted my attention from Fred's fossils in his palaeontology class… Reader, I married her.

(Adapted from *A Jurassic Romance*; *Geology* magazine, February 2016.)

## RHIAN JONES

*PhD, Manchester University, 1983–1986. Professor, Earth and Planetary Sciences, University of New Mexico, 1987– 2014. Reader in Petrology and Geochemistry of Meteorites, Manchester University, 2014 – present.*

I first came to the University of Manchester in 1983, as a new PhD student. I had done my Bachelor's degree in Chemistry at Oxford University, and slid into Geology sideways via a single undergraduate class in Mineralogy and a 4th-year undergraduate research project on silicate melts. I felt very fortunate that Dave Hamilton and Prof MacKenzie were willing to take me on to do an experimental petrology PhD project that did not require a traditional geology background. In those days, experimental petrology had a huge presence in the Williamson Building. It dominated the basement, with its rows of cold seal pressure vessels, piston cylinder instruments, and the internally heated pressure vessels in their protective steel rooms, which I used for my project. In addition, there was the large and busy workshop, the thin section facility, and analytical labs including XRD, SEM and electron microprobe. The probe was of course not automated, and I would settle in for long evenings of 2-minute analyses with my knitting. My office was in the basement, and I spent much of my days there, punctuated by trips upstairs to the tearoom. The tearoom was the hub of department life, with its daily cycles of buzzing conversations that ranged from the frivolous and banal to stimulating discussions of scientific ideas and collaborations. It was an egalitarian mix of academics, students, technical and administrative staff and we shared a common camaraderie. Social life continued at the Bowling Green following librarian Roy's weekly reminder, 'Don't forget it's Friday.' Being a chemist, my understanding of geology was gained mainly through osmosis, from tearoom conversations, weekly colloquia, and demonstrating for undergraduate courses. I tagged along on field trips to Anglesey, the Lake District, and the Isle of Mull which included a memorably beautiful day sailing to Staffa via Iona. Nobody seemed to care that I was neither student nor demonstrator, and they were invaluable trips for me in terms of my geology education. I had no idea that I would end up in the field of planetary science, but an encounter with the Apollo thin sections that were passing through the Department on loan made a profound impression on me. During the time I was a student, Prof Mac was busy taking photomicrographs for his beautiful Atlas volumes, and Prof Zussman was working on the second edition of the classic DHZ. The first conference I attended was the first European meeting devoted to experimental petrology, in Nancy, France. A group of us drove there in a minibus, which smelled rather ripe on our return on account of the large amounts of cheese we all brought back. I completed my PhD in 1986, after many late-night hours of drafting diagrams with ink pens and rub-on text and symbols. I felt very modern because I was able to type the text on a word processor. After my PhD I went with Adrian Brearley to Albuquerque, New Mexico, on what was supposed to be a two-year adventure as postdocs. Things turned out to be more long-term, as we got married there, had two children, and both established our careers in planetary science.

In 2014, I returned to the University of Manchester as a Reader, specializing in the petrology and geochemistry of meteorites. Times have certainly changed. We are now the much larger School of Earth and Environmental Sciences, and planetary science is a well-established component of undergraduate programmes as well as a thriving area of research. The experimental petrology lab has shrunk significantly, but the analytical facilities in the basement are still thriving and the microprobe is still reassuringly in the same space. Academic staff spend a large proportion of their days in front of a computer screen, and rarely visit the library. But plus ça change, and the Williamson Building still has the same wood panelling and the same huge wooden tables in the ground-floor teaching laboratories. The annual staff photos where I find my younger self as part of the group still adorn the first-floor corridor. A surprising number of people who I knew in the 1980s are still associated with the School. Shortly after I returned, I went to a retirement party for Giles Droop which was uncannily like returning to the 1980s tearoom, as it was attended by a remarkable number of the academic staff from my student days. From my perspective, the Department, now School, has maintained its traditional academic and teaching strengths, and is a place where learning and a love of the Earth continue to be respected and nurtured.

## ROGER MITCHELL

*Honours BSc, Manchester, 1964; MSc, Manchester, 1966. PhD, McMaster University, 1969; DSc, Manchester, 1978. Professor and latterly Emeritus Professor, Lakehead University, Thunder Bay, Canada.*

I attended Manchester University from 1961 to 1966 – an undergraduate student from 1961–64, and subsequently a graduate student, I was one of a first-year class of only 7 students who were some of the first undergrads to study in the new Williamson Building on Oxford Road.

My first memory of Manchester was the first lecture I attended. Dr G. D. Nicholls entered the lecture theatre, just to the left of building entrance, and proceeded to draw upon the blackboard the binary T-X phase diagram for the system forsterite – fayalite. This was to me, and my peers, mind-blowing, and it might as well have been in Sanskrit! Eventually, we learnt that this material was relevant to the crystallization of basalts – a rock type of which we were vaguely aware – and that we should go to 'The Library' – directly across on Oxford Road – and seek out the original work if we wished to understand the nuances of the subject! NO SPOON-FEEDING, THEN! HOW INTIMIDATING! Sort of like entering a cathedral – an old Victorian Gothic building – dark polished wood, stacks of books and journals (real library stacks), overseen by some stern and forbidding librarians – whose only function we thought was to keep us from using the books; totally unlike libraries of today. However, we did persevere and eventually knew where to find journals – journals such as *Norsk Geologisk Tidsskrift*; a real learning experience unfortunately not available to today's students since the advent of online resources.

A highlight of my first days – the senior students organized a field trip to Whitby and Scarborough for freshmen which gave us the opportunity to meet many fellow students prior to the start of the first term. This was a great experience; we got some idea of what we might be 'up against' in our future studies – we saw ammonites, fossil and (fresh) molluscs, enjoyed gin and orange for breakfast provided by a Whitby landlady, climbed the cliffs at Scarborough, and experienced the Khyber Pass! Overall it was what would nowadays be termed a 'bonding experience' – but then we called it camaraderie.

Field work organized by the Geology Club, together with the annual field school, eventually became very important for most of us both educationally and socially. Monthly excursions exposed us to a wide variety of fossils, rock types, and tectonic settings and was one of the highlights of my undergraduate times. However, many of us no doubt remember freezing in the unheated B&Bs of Dolgellau, getting soaked on the Derbyshire moors, and trying to find the Moine Thrust in the Scottish fogs!

As an undergraduate I learnt many strange things, such as the use of spherical trigonometry in crystallography (with Prof Zussman) and how to determine the pleochroic schemes of amphiboles (with Prof Howie); techniques which are now arcane and otiose, but in those times very relevant. I also came to appreciate the extensive knowledge of all of the faculty – and, on my part, especially Prof Mac (Prof W. S. MacKenzie) who kindled my interest in alkaline rock petrology. One of his favourite comments, in his broad Scottish brogue, when we were unable to identify some minor mineral in a thin section was – 'But, laddie, is it *important*?'! He had other sayings too about our observational abilities, but they are expurgated from this note.

Other outstanding memories: those of Prof R. A. Howie, at that time involved in publishing the first version of DH&Z. We were unsuspecting guinea pigs for this epic – although we did not realise it at the time. For example we were challenged, for our exams, to 'devise a classification of the amphiboles'. Dr John Dewey – I remember his unbridled enthusiasm in his work revolutionizing tectonic studies, although he was a classical structural geologist – a discipline for which some of us had little affinity, especially when we were instructed to observe the S3 cleavage in some Anglesey slates in the pouring rain! Dr Fred Broadhurst – from whom I gained my entire knowledge of invertebrate and – fortunately – vertebrate palaeontology; the former instilled in me to this day a knowledge of the zone fossils of Lias and the water vascular system of the echinoidea, and the latter an interest in skeletal architecture from an artistic viewpoint. All of this served me well, as one of my earlier teaching assignments – totally atypical for an igneous petrologist – was to teach undergraduate palaeo!

Manchester, during my time there, was one of the leading universities in the fields of geochemistry and experimental petrology. I found these to be the more fascinating and challenging areas of geoscience, and ultimately this led to a career as an igneous petrologist. At that time there were few textbooks, of which I managed to memorize all of Hatch, Wells & Wells and most of Turner & Verhoogen. The 1960s were the heyday of experimental petrology and I developed an 'unhealthy' interest in phase diagrams, especially those produced and propagated by Prof Mac and Dave Hamilton – using the mysterious and dangerous high-pressure equipment in the Williamson building basement – as these diagrams seemed to me to explain how magmas crystallized and evolved.

As a graduate student I learnt how to acquire geochemical data, interpret these data, and research topics for my MSc thesis. The skills I developed during that time have served me exceptionally well in my subsequent career. Unfortunately, some of the analytical techniques I used – spark source mass spectrometry (in which Manchester at that time was a world leader) and colorimetry – are now long forgotten, together with the mechanical calculators we used to 'compute' our results.

In addition to my university studies, Manchester provided me with several interesting social experiences not available in my home town area (Bradford & Leeds) such as: real bookshops (not the plethora of shoe shops which infested Leeds); the Hallé orchestra; the Moss Side slums; the real differences between Eire and Ulster; the Oxford Road cheap Chinese and Indian (not as good as Bradford) restaurants which kept us going after 3 hours of optical mineralogy; and some of the best green smog I have ever experienced.

## KENT BROOKS

*Honours BSc, Manchester University, 1961. DPhil, Oxford University. Employed as Professor, Institute of Geology, University of Copenhagen, 1968–2007, and subsequently Emeritus Professor.*

After 56 years, my recollections of undergraduate life in Manchester have become blurred by recollections from many parts of the world. However, two events spring to mind:

### Finding of a 'Prehistoric Monster'

At the time of the first event I was a second-year geology student on an excursion with Fred Broadhurst to the Yorkshire coast. I saw something pointed sticking out of the shales, took a blow at it with my hammer and said, 'Not another bloody belemnite.' It was picked up by the instructor (was he called Potts?), who was more observant than me. He said, 'This belemnite's got teeth.' We therefore started digging and quickly unearthed the skull and a couple of limb bones. The original find was the very tip of the jaws, where they came to a point. This was around Easter, and a party returned in the summer in a rented vehicle, equipped with appropriate digging gear, to collect the main part of the skeleton, although I didn't take part in this expedition.

At the time we could see it was big, but we didn't think as big as the local paper which described it as a *huge monster, 30 feet long*. It turned out to be a plesiosaur, probably from memory about 14 feet in length. It subsequently acquired the name 'Percy' and is a fossil of global significance. I was very fond of Fred – he almost persuaded me to become a palaeontologist! (Instead I have devoted my life to 'hard' rocks.)

### Immortality in a Mineral Name!

The second event only became significant in retrospect. It was in an exam set by Bob Howie. Bob got me fascinated with minerals, he made them come alive and ordered them so that they comprised a system which made them understandable: common minerals even took on the character of old friends. The final question in this exam was an unknown mineral (i.e. one which had not been studied in class). The object was to describe its properties, even though the student had not seen it before and was unable to name it. It turned out that my unknown mineral was eudialyte (everybody got something different), which I had decidedly never heard of before. Years later, eudialyte was recognized to be a rather extensive mineral group (to date at least 26 members, with additional members as yet unnamed). One member, rich in niobium and rare earths, was given the name kentbrooksite. It came from Kangerlussuaq in East Greenland (near to the Skaergaard Intrusion – made famous from the work of Professors Wager and Deer) and had originally been found by Professor Deer, who thought it to be eudialyte as it was prior to the proliferation of eudialyte varieties. I had collected the type specimen which was studied and named by Ole Johnsen of Copenhagen University and Canadian colleagues. Most new minerals nowadays are of microscopic size, but kentbrooksite is found as a rock-forming mineral in certain major intrusions. Further, the varieties carbokentbrooksite and ferrokentbrooksite have also been recognized. I would never have guessed this course of events when struggling under exam conditions to measure some of the unknown eudialyte's physical and optical properties all those years ago as a humble undergraduate in Manchester. Maybe if I had not passed this exam my career might have suffered, in which case there would not have been a mineral called kentbrooksite!

This is also a zirconium mineral and my DPhil work at Oxford was on the geochemistry of zirconium, of which I can confidently say that at the time I was a world expert – yet another reason why this mineral should bear my name!

## KATE BRODIE

*Lecturer and Senior Lecturer, University of Manchester, 1989–2017.*

I studied geology and geophysics at Leeds University but at the end of my first year decided geology was more interesting and transferred to a degree in Earth Sciences, followed by a PhD at Imperial College London, and then a lectureship. I joined the staff in Manchester in 1989, transferring with my husband and fellow geologist, Ernie Rutter, from Imperial as part of the 'Oxburgh Review'. Ernie brought his rock deformation laboratory to Manchester, which, with a well-established experimental petrology laboratory, analytical geochemical facilities and technical support, and a good Material Science Department with electron microscopy facilities, provided an excellent fit to our research interests.

Moving from Imperial, Manchester seemed a relatively small and friendly Department, despite the 50% increase in size at that time as a result of the geoscience review. In clearing my office as I have recently retired, I came across the 'mugshots' of the postgraduate students in my first year (1989/90): I am surprised that I remember them all! Times have changed since then as the Department has grown and expanded to include environmental and atmospheric scientists. My first recollection is painting my office – it had been occupied by a postgraduate student who smoked. Pam Collins, the sole finance person in those days, arranged to buy the paint. It was also my first encounter with Mac computers. Bill Sowerbutts had set up a computer lab for teaching with multiple Macs. Their graphical interface was so easy to use compared with the command lines which were required by PCs. Those computers lasted for years when PCs had to be regularly upgraded.

My son was very young and I had to bring him in one day when he was off school and we had to attend a first-aid course for field work, held in the large lecture theatre. He took it all in and had great fun practising bandaging up Bill Sowerbutts – this maybe gave him a taste for medicine as he is now a GP.

All the Department, academic, technical, secretarial and postgraduates, met for coffee in the tea room which was the central hub of the Department and where you found out what was going on. As the Department grew, and groups developed, separate tea areas formed but, despite a downsize to the current room, until recent years the geology staff still used to meet up for lunch, an important interaction for both research and teaching.

Freezing in the winter while teaching in the old Petrology Lab in the corner of Williamson above the library is another memory – now a geochemistry lab with hopefully improved heating. Early on I became involved in organizing lunchtime research seminars which was a great way of finding out about activities in the Department and getting to know research students. But even in those days, attendance was mixed and there were discussions about the best time for these to be held which continue to this day!

I have many happy memories of undergraduate field trips (as they were in those days, now field courses…) to Devon and Coniston initially, then Anglesey, NW Scotland and, in more recent years, France.

Photography was an important facility in the Department run by Sue Maher – making project slides, printed-out electron microscopy plate negatives, and large-format topographic field maps for field mapping. The analytical facilities and excellent technical support were a big attraction of our move to Manchester which has a long tradition of microprobe work. A JEOL SEM was acquired shortly after I arrived. The Cameca microprobe is still going strong, recently upgraded with new software, but the JEOL, after 30 years, is being dismantled as I retire, replaced by more cutting-edge equipment. And I am pleased to say, still with excellent technical support.

# CHAPTER 7

# OUTPUTS (AND SOME INPUTS) OF THE MANCHESTER DEPARTMENT AND RECOGNITION OF THE ACHIEVEMENTS OF MANCHESTER GEOLOGISTS

## TRAINED GEOLOGISTS, PUBLICATIONS, GRANTS AND AWARDS

The outputs of the Geology Department can be categorised under a number of headings, beginning firstly with trained 'geologists' or those taught some elements of geology as part of other degree programmes. This includes those taught to first (BSc) or higher degree (MSc, PhD) levels. Later years saw an extension of these traditional, usually three-year-long, degree programmes to include MPhil degrees and, more particularly, four-year undergraduate programmes called Masters in Earth Sciences (M. Earth Sci), or in Environmental Sciences for example. Tracking down accurate figures for the numbers of Manchester Geology graduates beginning in the mid 19th century is very difficult, although some indications are provided in Chapter 5. Our best estimates of the 'grand' total over the period from 1880 to 2003, outside of which the record is more difficult to discern, are five thousand geologists (broadly defined to include honours geologists and geochemists, for example) and another two thousand when the total includes students following combined honours or joint honours (two-subject) degree programmes such as geology with geography, or with chemistry, physics, or biology.

The second major outputs are the results of research undertaken in the department, sometimes in collaboration with the staff of other universities, institutes or laboratories. The great majority of such results have been published as articles in learned or technical journals, most of which are freely available in libraries as hard copies or electronically; some have been published via other media, most notably as geological maps, or, much more rarely, as patents. There are also confidential reports and other information most often associated with contracts with industry and other 'consulting' work. As with the difficulties of accurately estimating the grand total of Manchester Geology graduates, so determining the total number of publications carrying the name of the department is an impossible task. That said, an annual report on research in the department for the academic year 1999–2000 listed all publications for the calendar years 1999 and 2000, which came to fifty-two for 1999 and seventy-one for 2000. Tenured academic staff at this time numbered thirty, plus fifteen non-tenured Research Associates. Comparable figures for 2017 are eighty and fifty-five. If estimating the numbers of publications is not possible, actually listing those publications with their authors, titles, and journal details is similarly impossible. However, one category of published output which we can list in detail is published books (see Appendix B). They are appropriate in that they range from undergraduate textbooks to advanced research monographs, and so reflect the contributions of 'Manchester geologists' to research and also to teaching. Included in Appendix B are books written (or co-written) and edited (or co-edited), but not single chapters in books edited by others. First and subsequent editions are included so as to recognise the long-standing success of a volume, and that later editions often contain important new material.

Trained 'geologists' and published research are clear outputs of the Geology Department. Although not an output, rather more an 'input', but nevertheless an important indicator of performance, is the degree of success in attracting resources in terms of external funding (research grants, research studentships and research fellowships). Although some funds have been attracted from industry (oil and gas companies, mining, the nuclear industry) and others from the European Union, the great majority came from the research councils (PPARC, EPSRC and especially NERC). An estimate of grant holdings for 1999–2000 is a total in excess of £13,700,000.

## RECOGNITION OF MANCHESTER 'GEOLOGISTS'

The honours and awards received by geologists (broadly defined) who have worked at Manchester University have been tangible recognition of their contributions to science. These honours and awards have ranged from knighthoods and honorary degrees, to Fellowships of the Royal Society and other learned bodies, to medals awarded by those bodies or election to the prestigious role of President. Some learned societies promote the best work in their subjects by inviting researchers to take on the role of 'Distinguished Lecturer' (normally this involves giving lectures via a tour of universities or institutions such as geological surveys). A unique form of recognition for geologists is having a newly discovered mineral named after them.

Minerals named for 'Manchester scientists' are listed in Appendix A along with information on the broad type of mineral concerned and its chemical formula. There are in fact eight such minerals; four are silicates, one is a sulphide, and the other three alloys or compounds of precious metals (e.g. silver, platinum, palladium).

The major honours and awards have been as follows:

- **Knighthood:** Boyd Dawkins, Holland, Pugh, Miers
- **Damehood (DBE):** Strank
- **Officer of the British Empire (OBE):** Curtis
- **Honorary Doctorate (Athens University):** Zussman
- **Honorary Doctorate (Derby University):** Howie
- **Honorary Member of the Mineralogical Society of Poland:** Zussman
- **Fellow of the Royal Society:** Williamson, Boyd Dawkins, T. H. Holland, Jones, Pugh, Hickling, D. M. S. Watson, Miers, Nockolds, Dewey, Deer, Fyfe, Turner, Wood
- **(Foreign) Fellow of the Royal Society of Canada:** Vaughan
- **Fellow of the Royal Society of Edinburgh:** MacKenzie, T. H. Holland
- **Fellow of the American Geophysical Union (AGU):** Rutter
- **Geochemical Fellow of the Geochemical Society:** Vaughan
- **Fellow of the Society of Economic Geologists:** Vaughan
- **Fellow of the Mineralogical Society of America:** Vaughan, Zussman, Henderson, Fyfe, Wood, Deer, Howie, Redfern, A. B. Thompson
- **Honorary Life Fellow of the Mineralogical Society of Great Britain and Ireland:** Brodie, Vaughan
- **Royal Astronomical Society Gold Medal:** Turner
- **President of the Geological Society of London (GSL):** T. H. Holland, Jones, Deer, Curtis, Manning, C. Holland
- **President of the Mineralogical Society of Great Britain and Ireland (MinSoc):** Miers, T. H. Holland,

Deer, Jones, Vincent, Howie, Zussman, MacKenzie,
Vaughan, Henderson, Pattrick

- **President of the Mineralogical Society of America (MSA):** Vaughan
- **President of the European Mineralogical Union (EMU):** Vaughan
- **President of the Geology Section, British Association for the Advancement of Science:** Pugh, T. H. Holland, Vaughan
- **President of the Institute of Mining and Engineering, London:** T. H. Holland
- **Royal Society Rumford Medal:** Turner
- **Royal Society Royal Medal:** Williamson, Jones
- **Royal Society Bakerian Lecture:** Williamson
- **Meteoritical Society Leonard Medal:** Turner
- **Royal Society of Chemistry Medal and Award for Geochemistry:** Vaughan
- **Royal Meteorological Society Gold Medal:** Choularton
- **GSL Murchison Medal:** Pugh, Hickling, Nockolds, Deer, Howie, Wood, Curtis
- **GSL Wollaston Medal:** Williamson, Jones, Miers, Dewey, Fyfe
- **GSL Lyell Medal:** Boyd Dawkins, Jones, Dewey, Rutter
- **GSL Wollaston Fund:** Jones, Vincent, Manning, Rutter
- **GSL Lyell Fund:** Nockolds, Gawthorpe
- **GSL Murchison Fund:** T. H. Holland, Tyrell, Deer, Dewey
- **GSL Bigsby Medal:** T. H. Holland, Ballentine, Lloyd
- **GSL Coke Medal:** C. H. Holland
- **GSL Prestwich Medal:** Boyd Dawkins

- **GSL William Smith Fund:** Flint
- **GSL President's Award:** Pawley, Brocklehurst
- **Palaeontological Association Lapworth Medal:** C. H. Holland
- **Alexander von Humboldt Research Award:** Selden
- **MinSoc Schlumberger Medal:** Henderson, Vaughan, Wood, S. Redfern, D.Manning, Lloyd
- **MinSoc Max Hey Medal:** Pawley, S. Redfern, Warren, Coker
- **MSA Award:** Wood, Fyfe
- **MSA Public Service Award:** Howie
- **MSA Roebling Medal:** Wood, Fyfe
- **European Association of Geochemists Urey Medal:** Turner
- **European Geoscience Union Neel Medal:** Rutter
- **Geochemical Society Goldschmidt Award:** Wood
- **MinSoc Hallimond Lecturer:** Zussman, Agrell, Hudson Edwards, MacKenzie, Turner, Wood
- **MSA Distinguished Lecturer:** Vaughan
- **MinSoc Distinguished Lecturer:** Vaughan, Manning, Morris
- **MinSoc Special Issue of *Mineralogical Magazine* in Honour:** Deer, Howie, Zussman
- **American Association of Petroleum Geologists (AAPG) Distinguished Lecturer:**
  - Gawthorpe
- **AAPG Wallace Pratt Distinguished Lecturer:** MacQuaker

The above list must inevitably be incomplete, particularly in regard to such recognition as afforded by honorary degrees. We apologise for such omissions and welcome any additions or amendments to this list.

It is worth recording that current Manchester Geology students can also achieve recognition through a variety of awards, either endowed in the memory of former staff of the department (W. Boyd Dawkins, MacKenzie and Guilford, F. M. Broadhurst) or in memory of former students, particularly those whose lives have been sadly cut short (W. H. Morton, Angela Speakman, Calvert Armstrong, Colin Hatfield). In addition to these endowments, there are awards made by national or international learned societies to recognise outstanding performance. These include the Mineralogical Society of Great Britain and Ireland, the Palaeontological Association, the Petroleum Exploration Society of Great Britain, and Petros Geosciences. The department is grateful to all of the above for the generous support given to its students over the years.

# CHAPTER 8

# EPILOGUE: THE YEARS AFTER 2004

In the autumn of 2004, the Victoria University of Manchester (VUM) merged with the University of Manchester Institute of Science and Technology (UMIST) to form a new University of Manchester. With the merging of the Physics Departments of the two universities, it was decided that a group of UMIST atmospheric physicists would be best served by joining our department to form a new School of Earth, Atmospheric and Environmental Sciences. The Atmospheric Physics Group at UMIST had been founded in 1961 by Professor John Latham, who moved from Imperial College. This merger entailed the relocation of nine established academic staff along with associated laboratories, staff offices, support staff, research staff and students to space in the Simon Building, opposite the Williamson Building on Brunswick Street.

At the time of writing (2017), there are fifteen academic staff in Atmospheric Sciences: Professors Choularton, Coe, Gallagher, McFiggans, Schultz, G. Vaughan and Webb, and Drs Alfarra, Allan, Allen, Bower, Connolly, Crosier, Garcia-Carreras and Topping. In addition the centre hosts eight research scientists from the NERC National Centre for Atmospheric Sciences (NCAS). Emeritus Professors Latham and Jonas, both ex-Heads of the Department of Physics at UMIST, remain closely associated with the school. Associated with this merger, the new university agreed to take on Professor Geraint Vaughan, a distinguished atmospheric physicist from Aberystwyth University who later was also appointed the NCAS Director of Observations, an appointment he holds alongside the Manchester Chair. The atmospheric physicists brought with them substantial resources, particularly including research aircraft operated by the group (see Figure 19).

As well as obviously offering substantial new areas of science to strengthen the environmental portfolio of the new school, there were the benefits of new collaborations involving Manchester 'geologists'. Examples of this synergy included work on atmospheric mineral dusts and other particulates collected both at high altitudes and on the large scale by research aircraft, and at ground or near-ground level in cities such as Manchester. One example of the former was work by Hugh Coe, David Vaughan, their students and postdocs on the mineralogy of dusts carried from desert soils in areas in West Africa and their interaction with sooty material produced by biomass burning. This has led to the development of single particle mass spectrometry to characterise desert dusts and examine their mineralogy, and of the mineralogy expertise in Manchester to aid the development of the use of UK synchrotron facilities to examine atmospheric particulates for the first time. Most recently, Professor Gallagher's expertise in measurement of biological material in the atmosphere by fluorescence has been combined with Professor Lloyd's leadership in geomicrobiology to examine the way viruses, bacteria and pollen behave in the atmosphere. As well as the new research (and teaching) opportunities provided by the acquisition of the UMIST atmospheric physicists, two of the Professors from the group served with distinction as Heads of the School (Department): Tom Choularton (2006–2009) and Hugh Coe (2009–2012). Tom had already served as Head of the UMIST Physics Department for two years leading up to the merger. He also has the distinction of being awarded the Mason Gold Medal by the Royal Meteorological Society for outstanding contributions to atmospheric science. Hugh's research focus is on improving our knowledge of the physics and chemistry of atmospheric aerosols which can play key roles in both climate change and air pollution.

At the time of writing (2017), the school has just undergone another dramatic change in size and character with the arrival of sixteen established academic staff from the biosciences. This was a result of the university deciding to break up its Faculty of Life Sciences, with the majority of the staff joining schools in the Medical Faculty, and those biologists working in the environmental biosciences joining the geoscientists. The staff joining us were Professors Bardgett, Brown, Chamberlain, D. Johnson and

Sellers, and Drs Buckley, De Vries, Gilman, G. Johnson, Knight, Pittman, Sansom, Semchenko, S. Schultz, Walton and White. As well as strengthening the new school for the teaching of many aspects of environmental sciences, this development offers great opportunities for further collaborative research ventures, whether in fundamental areas such as molecular environmental science or studies of ancient life (including dinosaurs!), or applied areas such as soils. This latest change has led to another change of name for what was long called the department, and which is now the School of Earth and Environmental Sciences. In spite of the staff changes associated with the arrival of large numbers of atmospheric physicists and bioscientists, it is very pleasing to note that there have been important staff developments in the core areas of geology, mineralogy and geochemistry. These have been partly associated with retirements (and 'semi-retirements') of staff in these core areas, and also with new positions. Appointments made after 2004 (along with their general fields of expertise) include Mike Burton (Chair in Volcanology), Patricia Clay (Research Fellow in Isotope Group), Vicky Coker (Lecturer in Environmental Mineralogy), Sarah Crowther (Research Fellow in Isotope Group), Steve Flint (Professor of Stratigraphy), Russell Garwood (Lecturer in Earth Sciences), Margaret Hartley (Lecturer in Earth Sciences), Greg Holland (Senior Lecturer in Terrestrial Noble Gas Geochemistry), Cathy Hollis (Reader in Petrophysics and Production Geoscience), Mads Huuse (Professor in Basin Studies), Rhodri Jerrett (Lecturer in Geology), Rhian Jones (Reader in Isotope Geochemistry and Cosmochemistry), Katherine Joy (Royal Society Research Fellow and Reader in Isotope Group), Ian Kane (Reader in Sedimentology), Phillip Manning (Professor of Natural History, and STFC Science in Society Fellow), Julian Mecklenburgh (Lecturer in Structural Geology), Kath Morris (BNFL Research Chair in the Geological Disposal of Radwaste), Brian O'Driscoll (Senior Lecturer in Petrology), Margherita Polacci (Senior Research Fellow), Laura Richards (Leverhulme Early Career Research Fellow), Clare Robinson (Senior Lecturer working on fungi), Stefan Schroeder (Senior Lecturer in Basin Studies), Sam Shaw (Professor in Environmental Mineralogy), Romaine Tartese (Research Fellow) and Bart Van Dongen (Reader in Organic Geochemistry).

A further appointment in the geosciences (including petroleum geology, but also active in other areas) was Kevin Taylor. Kevin had spent time in the department working with Charles Curtis in the early '90s, and in later years worked at the Manchester Metropolitan University. In 2015 he took over from Hugh Coe as Head of the School of Earth and Environmental Sciences. So, once again, the Head of 'Department' is a geologist under whose leadership geology at Manchester University goes from strength to strength; from just Williamson and his helper Boyd Dawkins, to a school with an academic staff complement of eighty and around fifty-five postdoctoral Research Associates.

# APPENDIX A

# MINERALS NAMED IN HONOUR OF MANCHESTER UNIVERSITY GEOLOGISTS

| Named | After | Mineral Description | Formula | Named By |
|---|---|---|---|---|
| Agrellite | S. O. Agrell | (Na,Ca) silicate | $NaCa_2Si_4O_{10}F$ | J. Gittins et al. |
| Deerite | W. A. Deer | (Fe,Mn,Al) silicate | $(Fe^{2+},Mn)_6(Fe^{3+},Al)_3Si_6O_{17}O_3(OH)_5$ | S. O. Agrell |
| Eugenite | E. F. Stumpfl | Silver amalgam | $Ag_9Hg_2$ | H. Kucha |
| Howieite | R. A. Howie | (Na,Fe,Mn,Al) silicate | $Na(Fe^{2+},Mn)_{10}(Fe^{3+},Al)_2Si_{12}O_{31}(OH)_{13}$ | S. O. Agrell |
| Stumpflite | E. F. Stumpfl | Platinum/antimony | $Pt\ Sb$ | Z. Johan and P. Picot |
| Vaughanite | D. J. Vaughan | (Tl,Hg,Sb) sulphide | $Tl\ HgSb_4S_7$ | D. C. Harris and A. C. Roberts |
| Vincentite | E. A. Vincent | Palladium/platinum | $(Pd,Pt)_3(As,Sb,Te)$ | E. F. Stumpfl and M. Tarkian |
| Zussmanite | J. Zussman | (K,Fe,Mg,Mn)Al silicate | $K(Fe^{2+}Mg,Mn^{2+})_{13}(Si,Al)_{18}O_{42}(OH)_{14}$ | S. O. Agrell |

# APPENDIX B

## BOOKS PUBLISHED BY MANCHESTER UNIVERSITY GEOLOGISTS (IN ORDER OF PUBLICATION YEAR)

**Boyd Dawkins, W.** 1874. *Cave Hunting*. Macmillan & Co. 442 pp.

**Boyd Dawkins, W.** 1880. *Early Man in Britain and His Place in the Tertiary Period*. Macmillan & Co. 537 pp.

**Williamson, W. C.** 1896. *Reminiscences of a Yorkshire Naturalist*. George Redway, London. 228 pp. (Facsimile edition published by J. Watson & B. A. Thomas, 1985.)

**Miers, H. A.** 1902. *Mineralogy: An Introduction to the Scientific Study of Minerals*. Macmillan & Co. London and New York. 620 pp.

**Hickling, G.** 1910. *Geology: Chapters of Earth History.* Frederick A. Spores Ltd. 154 pp. (In later years, this book was produced by other publishers; see https://amazon.uk.com.)

**Stopes, M. C.** 1910. *Ancient Plants.* Blackie & Son Ltd., London. 290 pp. (Later copies; see above.)

**Stopes, M. C.** 1912. *Botany, or the Modern Study of Plants.* T. C. & E. C. Jack Ltd., London and Edinburgh.100 pp. (Also see Nelson & Sons Ltd.)

**Jones, O. T.** 1922. *Lead and Zinc: The Mining District of North Cardiganshire and West Montgomeryshire.* Special reports on the mining reserves of Great Britain, Vol. XX. The Geological Survey, His Majesty's Stationery Office. 207 pp.

**Wager, L. R. and Deer, W. A.** 1939. *Geological Investigations in East Greenland, Part III: The Petrology of the Skaergaard Intrusion, Kangerdlugssuaq, East Greenland.* Meddel. Gron. Vol.105, 1– 352.

**Buckley, H. E.** 1951. *Crystal Growth.* Chapman & Hall, London; John Wiley, New York. 571 pp.

**Charlesworth, J. K.** 1953. *The Geology of Ireland: An Introduction.* Oliver & Boyd, Edinburgh. 276 pp.

**Fyfe, W. S.**, Turner, F. J. and Verhoogen, J. 1958. *Metamorphic Reactions and Metamorphic Facies; Geological Society of America Memoir.* Vol.73, 259 pp.

**Deer, W. A., Howie, R. A. and Zussman, J.** *Rock-Forming Minerals*:

1962. Vol. 1, *Chain Silicates.* Longmans, Green & Co Ltd., London, 333 pp.

1962. Vol. 3, *Sheet Silicates.* Longmans,Green & Co. Ltd., London. 270 pp.

1962. Vol. 5, *Non-Silicates.* Longmans, Green & Co. Ltd.,London. 371 pp.

1963. Vol. 2, *Chain Silicates*. Longmans, Green & Co. Ltd., London. 379 pp.

1963. Vol. 4, *Framework Silicates*. Longmans, Green & Co. Ltd., London. 435 pp.

**Fyfe, W. S.** 1964. *Geochemistry of Solids*. McGraw-Hill, New York, 199 pp.

**Deer, W. A., Howie, R. A. and Zussman, J.** 1966. *An Introduction to the Rock-Forming Minerals*. Longmans, Green & Co., London. 528 pp.

**Zussman, J.** (Editor). 1967. *Physical Methods in Determinative Mineralogy*. Academic Press, London. 514 pp.

**Broadhurst, F. M.**, Eager, R. M. C., Jackson, J. W., **Simpson, I. M.** and Thompson, D. B. 1970. *The Area Around Manchester*. The Geologists' Association. 118 pp.

**Fyfe, W. S.** 1974. *Geochemistry*. Oxford University Press, Oxford. 120 pp.

**Simpson, I. M. and Broadhurst, F. M.** 1975. *A Building Stones Guide to Central Manchester*. Manchester Geological Association. 39 pp.

**Elder, J. W.** 1976. *The Bowels of the Earth*. Oxford University Press, Oxford. 222 pp.

**Wood, B. J.** and Fraser, D. G. 1976. *Elementary Thermodynamics for Geologists*. Oxford University Press, Oxford. 303 pp.

**Zussman, J.** (Editor). 1977. *Physical Methods in Determinative Mineralogy*. Second Edition. Academic Press, London. 720 pp.

**Deer, W. A., Howie, R. A. and Zussman, J.** 1978. *Rock-Forming Minerals*. Second Edition, Vol. 2A, *Single-Chain Silicates*. Longman Group UK Ltd, Harlow, Essex. 668 pp.

**Nockolds, S. R.**, Knox, R. W. O. and Chinner, G. A. 1978. *Petrology for Students*. Cambridge University Press, Cambridge. 443 pp.

**Vaughan, D. J.** and Craig, J. R. 1978. *Mineral Chemistry of Metal Sulfides*. Cambridge University Press, Earth Science Series, Cambridge. 500 pp.

**MacKenzie, W. S. and Guilford, C.** 1980. *Atlas of Rock-Forming Minerals in Thin Section*. Longman Group UK Ltd, Harlow, Essex. 98 pp.

Craig, J. R. and **Vaughan, D. J.** 1981. *Ore Microscopy and Ore Petrography*. Wiley Interscience, New York. 406 pp.

**Elder, J. W.** 1981. *Geothermal Systems*. Academic Press, London, 508 pp.

**Holland, C. H.** 1981. *A Geology of Ireland*. Scottish Academic Press, Edinburgh. 335 pp.

Newton, R. C., Navrotsky, A. and **Wood, B. J.** (Editors). 1981. *Thermodynamics of Minerals and Melts*. Springer e-book.

**MacKenzie, W. S., Donaldson, C. H. and Guilford, C.** 1982. *Atlas of Igneous Rocks and Their Textures*. Longman Group UK Ltd, Harlow, Essex. 148 pp.

**Vaughan, D. J.** (Guest Editor). 1983. *Sulphide Mineralogy and Petrology with Special Reference to Metamorphic Rocks; Mineralogical Magazine*, Vol. 47, Part 4.

**Adams, A. E., MacKenzie, W. S. and Guilford, C.** 1984. *Atlas of Sedimentary Rocks Under the Microscope*. Longman Group UK Ltd, Harlow, Essex. 104 pp.

Berry, F. J. and **Vaughan, D. J.** (Editors). 1985. *Chemical Bonding and Spectroscopy in Mineral Chemistry*. Chapman & Hall, London. 325 pp.

Walther, J. V. and **Wood, B. J.** (Editors). 1986. *Fluid Rock Interactions During Metamorphism*. Springer Verlag, New York. 211 pp.

**Elder, J. W.** 1987. *The Structure of the Planets*. Academic Press, London. 210 pp.

Craig, J. R., **Vaughan, D. J.** and Skinner, B. J. 1988. *Resources of the Earth: Origin, Use and Environmental Impact*. Prentice Hall, New Jersey. 385 pp.

Holloway, J. R. and **Wood, B. J.** 1988. *Simulating the Earth: Experimental Geochemistry*. Springer e-book. 196 pp.

**Vaughan, D. J.** (Guest Editor). 1989. *Spectroscopic Studies of Minerals: Principles, Applications and Advances*; *Mineralogical Magazine*, Vol. 53, Part 2.

Jambor, J. L. and **Vaughan, D. J.** (Editors). 1990. *Advances in Microscopic Study of Ore Minerals*. Mineralogical Association of Canada, Ottawa. 426 pp.

Yardley, B. W. D., **MacKenzie, W. S.** and **Guilford, C.** 1990. *Atlas of Metamorphic Rocks and Their Textures*. Longman Group UK Ltd, Harlow, Essex. 120 pp.

Eager, R. M. C. and **Broadhurst, F. M.** 1991. *Geology of the Manchester Area*. Second Edition. The Geologists' Association. 118 pp.

**Kerrick, D. M.** 1991. *Contact Metamorphism*; *Reviews in Mineralogy*, Vol. 26. Mineralogical Society of America. 672 pp.

**Deer, W. A., Howie, R. A.,** and **Zussman J.** 1992. *An Introduction to Rock-Forming Minerals*, Second Edition, Longman Group UK Ltd, Harlow Essex. 696 pp.

Tossell J. A. and **Vaughan. D. J.** 1992. *Theoretical Geochemistry: Applications of Quantum Mechanics in the Earth and Mineral Sciences*. Oxford University Press, Oxford and New York. 514 pp.

**Treagus. J. E.** 1992. (Editor). *Caledonian Structures in Britain: South of the Midland Valley*. Chapman and Hall, London. 196 pp.

**Vaughan, D. J.** 1993. (Editor, Mineralogy Section). *The Encyclopaedia of the Solid Earth Sciences*. Blackwells, Oxford, 713 pp,

**Manning, D. A. C.** 1994. *Introduction to Industrial Minerals.* Chapman & Hall, London. 276 pp.

Coleman, M. L., **Curtis, C. D.** and **Turner, G.** (Editors). 1994. *Quantifying Sedimentary Geochemical Processes.* Royal Society. 186 pp.

**MacKenzie, W. S. and Adams, A. E.** 1994. *A Colour Atlas of Rocks and Minerals in Thin Section.* Manson Publishing Ltd, and John Wiley & Sons, USA. 192 pp.

Craig, J. R. and **Vaughan, D. J.** 1994. *Ore Microscopy and Ore Petrography.* Second Edition. Wiley Interscience, New York. 368 pp.

**Vincent, E. A.** 1994. *Geology and Mineralogy at Oxford 1860–1986: History and Reminiscence.* Department of Earth Sciences, University of Oxford. 245 pp.

**Vaughan, D. J. and Pattrick, R. A. D.** (Editors).1995. *Mineral Surfaces.* Chapman & Hall, London. 370 pp.

Chang, L. L. Y., **Howie, R. A.** and **Zussman, J.** 1996. *Rock-Forming Minerals.* Second Edition, Vol. 5B, *Non-Silicates (Sulphates, Carbonates, Phosphates and Halides).* Longman Group UK Ltd, Harlow, Essex, 383 pp.

Craig, J. R., **Vaughan, D. J.** and Skinner, B. J. 1996. *Resources of the Earth: Origin, Use and Environmental Impact.* Second Edition. Prentice Hall, New Jersey. 472 pp.

**Deer, W. A., Howie, R. A. and Zussman, J.,** *Rock-Forming Minerals,* Second Edition:

Vol. 1A, 1997. *Orthosilicates.* The Geological Society of London. 919 pp.

Vol. 1B, 1997. *Disilicates and Ring Silicates.* The Geological Society of London. 629 pp.

Vol. 2B, 1997. *Double-Chain Silicates.* The Geological Society of London. 764 pp.

Cabri, L. J. and **Vaughan. D. J.** 1998. *Modern Approaches to Ore and Environmental Mineralogy.* Mineralogical Association of Canada, Ottawa. 434 pp.

**Vaughan, D. J.** and **Wogelius, R. A.** (Co-Editors). 2000. *Environmental Mineralogy; Notes in Mineralogy.* European Mineralogical Union. 434 pp.

Craig, J. R., **Vaughan, D. J.** and Skinner, B. J. 2001. *Resources of the Earth: Origin, Use and Environmental Impact.* Third Edition. Prentice Hall, New Jersey. 385 pp.

**Fleet, M. E.** 2003. *Rock-Forming Minerals.* Second Edition. Vol. 3A, *Micas.* The Geological Society of London. 758 pp.

**Deer, W. A., Howie, R. A.,** Wise, W. S. and **Zussman, J.** 2004. *Rock-Forming Minerals.* Second Edition. Vol. 4B, *Framework Silicates (Silica Minerals, Feldspathoids and the Zeolites).* The Geological Society of London. 982 pp.

**Vaughan, D. J.** and Jambor, J. L. (Guest Editors). 2006. *Mineralogy and Geochemistry of Acid Mine Drainage and Metalliferous Minewastes; Applied Geochemistry,* Vol.21, No. 8.

**Vaughan, D. J.** (Guest Editor). 2006. *Arsenic; Elements,* Vol. 2, No. 2. 63 pp.

**Vaughan, D. J.** (Editor). 2006. *Sulphide Mineralogy and Geochemistry; Reviews in Mineralogy and Geochemistry.* Geochemical Society and Mineralogical Society of America, Vol.61. 714 pp.

**Simpson, I. M.** and **Broadhurst, F. M.** 2008. *A Building Stones Guide to Central Manchester.* Second Edition. Manchester Geological Association. 45 pp.

**Vaughan, D. J.** and Thompson, A. (Guest Editors). 2009. *Geochemistry and Mineralogy of Metalliferous Mine-Wastes: An Issue in Honour of John Jambor; Applied Geochemistry,* Vol. 24. 162 pp.

Craig, J. R., **Vaughan, D. J.** and Skinner, B. J. 2011. *Earth Resources and the Environment.* (Formerly *Resources of the Earth: Origin, Use and Environmental Impact.*) Fourth Edition. Prentice Hall, New Jersey. 508 pp.

Bowles, J. F. W., **Howie, R. A., Vaughan, D. J.** and **Zussman, J.** 2011. *Rock-Forming Minerals. Non-Silicates (Oxides, Hydroxides and Sulphides).* Second Edition. The Geological Society of London. 920 pp.

**Fyfe, W. S., Thompson, A. B.** and Price, N. J. 2012. *Fluids in the Earth's Crust. Their Significance in Metamorphic, Tectonic and Transport Processes.* Elsevier, 402 pp.

**Deer, W.A., Howie, R.A.** and **Zussman, J.** 2013. *Introduction to the Rock-Forming Minerals* (Third Edition). Mineralogical Society, London. 498 pp. + Crystal Viewer Structures Compact Disc.

**Treagus, J. E.** and **Treagus, S. H.** 2013. *The Rocks of Anglesey's Coast.* Llygad Gwalch. Cyf ,192 pp.

**Vaughan, D. J.** and **Wogelius, R. A.** (Editors). 2013. *Environmental Mineralogy II. EMU Notes in Mineralogy.* European Mineralogical Union. Vol.13, 489 pp.

**Vaughan, D. J.** 2014. *Minerals: A Very Short Introduction.* Oxford University Press, Oxford. 140pp. (Invited contribution for a general audience.)

# TABLE 1

# UNDERGRADUATE NUMBERS

Total (first + second + third year) numbers per calendar year (averaged for each decade)

| Decade | Geology | Geochemistry | Environment and Resources | Combined Honours | Geology and Mining | M. Science |
|---|---|---|---|---|---|---|
| 1880s* | 20 | — | — | — | — | — |
| 1890s | 12 | — | — | — | — | — |
| 1900s | 8 | — | — | — | 9 | — |
| 1910s | 10 | — | — | — | 6 | — |
| 1920s | 12 | — | — | — | 3 | — |
| 1930s | 15 | — | — | — | 8 | — |
| 1940s | 27 | — | — | — | — | — |
| 1950s | 36 | — | — | — | — | — |
| 1960s | 48 | — | — | 31 | — | — |
| 1970s | 80 | — | — | 50 | — | — |
| 1980s | 85 | — | — | 58 | — | — |
| 1990s | 142 | 50 | 51 | 86 | — | — |
| 2000–2003 | ^139 | 10 | 16 | 88 | — | **13 |

**Totals:** From 1880 to 2003, about 5,000 single honours and 2,000 combined honours students.

\* May have included some first-year students taking geology only as a subsidiary course.

^ Averaged over four years.

\*\* Started 2002, so averaged over two years.